Limnology of the Red Lake, Romania

Gheorghe Romanescu
Cristian Constantin Stoleriu
Andrei Enea

Limnology of the Red Lake, Romania

An Interdisciplinary Study

Springer

Gheorghe Romanescu
Cristian Constantin Stoleriu
Andrei Enea
Faculty of Geography and Geology
Department of Geography
University "Alexandru Ioan Cuza"
Iaşi
Romania

ISBN 978-94-007-6756-0 ISBN 978-94-007-6757-7 (eBook)
DOI 10.1007/978-94-007-6757-7
Springer Dordrecht Heidelberg New York London

Library of Congress Control Number: 2013935772

Printed on acid-free paper

Springer is part of Springer Science+Business Media (www.springer.com)

Preface

Limnology is the border discipline claimed by biology and geography. This is the reason why limnobiology or limnogeography has an ever-increasing use in the field. Though limnology is interdisciplinary by definition, it led to the creation of specializations such as limnogeology, limnogeography, limnoarchaeology, etc.

In this case, we present an interdisciplinary approach based on underlying the geomorphologic characteristics. From this perspective, the authors can definitely say that the approach is new and that it emphasizes the role played by the geomorphologic factors in the distribution and combination pattern of other limnological factors.

The graphics used are exceptional and some representations are new in the field: the map of the lacustrine basin cliffs; the map illustrating the orientation of slopes within the lacustrine basin; the delimitation of wetlands through the correlation between depth, cliffs, and orientation of the slopes; the graphic representation of aquatic bodies, on the vertical, with the corresponding water volume for each level, etc. All these characteristics make this book a statement in the scientific literature of the field.

The existence of numerous lakes, with various origins, makes the Eastern Carpathians an area of main interest for Romanian tourism. The mountain climate—with Baltic and ocean influences, where temperatures are moderate and precipitations are abundant (700–1,400 mm)—favors the existence of lacustrine basins and wetlands (Apăvăloaie 1971).

The Red Lake, through its genesis and landscape, represents one of the most important tourist attractions in Eastern Romania. It is also favored by the existence of a modernized access road, connecting Transylvania and Moldavia. This is another reason why a relevant study was needed to underline its basic features, its national and international individuality, and to spread the information globally.

Most of the corresponding lakes and wetlands are affected by anthropic interventions but also by non-intervention. The decrease in the lacustrine surfaces, as a consequence of normal clogging, is a natural phenomenon. When erosion is accelerated (anthropic intervention), the depth is reduced and, most of the time, the surface increases. For the Red Lake, such a scenario is not possible because the waters in excess are eliminated over the top of a natural dam, towards the Bicaz River. Through clogging, the surface occupied by wetlands increases but the aquatic surface decreases. The ecologic dichotomy, of maintaining the water surface, or of extending wetlands, is visible.

Wetlands, for the entire planet, have been deeply altered, in different manners, both in space and time. The wetlands and deep waters within the mountain units of the Eastern Carpathians are in an incipient stage of human intervention. For the entire country, the morphometric characteristics and the climatic conditions allowed the formation of a wide variety of wetlands; their preservation was supported by a few possibilities of drainage or by low-habitation density.

In the Eastern Carpathians, research underlines more and more the pollution of streams and of the lacustrine basins, implicitly. As a consequence of low altitudes and of the existence of numerous passes and gorges, the Eastern Carpathians represents an important habitation area, where the population density is relatively high. This has led to the alteration of streams by the building of barrages and dams and by the elimination of floodplains, etc.

The most important argument for the preservation and rehabilitation of lacustrine basins and wetlands is represented by their complex role: regulating the hydrologic cycle; conserving the biodiversity; reducing soil erosion; improving the microclimate; reducing high-water waves; tourist destination, etc.

The lifespan of natural dam lakes has been reduced. This is the reason why these water units attract a large number of tourists. The management of these basins is extremely delicate and an alteration that could seem to bring benefits may lead inevitably to the degradation of other characteristics. In consequence, human intervention is not welcome as the waters of such lakes could be lost.

Ioan Nistor
University of Ottawa
Canada

Acknowledgments

The true limnological and limno-archaeological studies within Alexandru Ioan Cuza University in Iaşi started when the Interdisciplinary Research Platform ARHEOINVEST was founded in 2006. In this sense, the authors thank Professor Victor Spinei, Ph.D., corresponding member of the Romanian Academy, the director of this platform, and lecturer Vasile Cotiugă, Ph.D., the administrative manager and the initiator of this idea.

All the equipment used during the data collection process belong to the Geoarcheology Laboratory within the Faculty of Geography and Geology, the Department of Geography (manager—Professor Gheorghe Romanescu, Ph.D.).

The project was financed by the Ministry of Education in Romania, through CNCSIS Grant No. 426 (2007–2009) and from our own resources.

For the organic carbon analyses and for certain clarifications regarding the vegetal associations of the wetlands related to the Red Lake, we thank associate professor Angela Lupaşcu, Ph.D., biogeographer within the Faculty of Geography and Geology, the Department of Geography.

The authors also thank the three referents who contributed significantly to the consolidation of this scientific endeavor and to revising data from the international literature: Professor Andre Ozer, Ph.D. (Belgium), Professor Ioan Nistor, Ottawa, Ph.D. (Canada), Professor Simon Hutchinson, Ph.D. (England).

We also thank the administration of the Bicaz Gorges Red Lake National Park for their logistic support.

Contents

Introduction

Limnology, as a border science, has long been avoided by the disciplines that are supposed to study the understanding of interdependent phenomena: geography and biology. Unfortunately, even at the international level, the phenomena are not acknowledged due to the lack of main components: the studies are either purely biological or purely geographical.

Most biological studies do not take into account the physical-geographic factors. At the same time, the geographic studies lack the biological component. Those who try to uniform the components get, surprisingly, an elegant rejection: "we regret it, but the subject chosen is not among the themes promoted by our journal". We cannot fully understand the role of each component and its connections with the whole. In limnology, there has always been a "missing link".

The schools in the field fight fiercely in order to impose their perspective in renowned specialties. How many international journals, well known in the field, accept articles with complex, interdisciplinary subjects? How many biologists know what a hydrographic basin is or what its geological constitution comprises? How many specialists take into account the morpho-bathymetric phase as a layering unit for local conditions? The lithological—emersed or submerged—substrate is the development support for the communities on various evolution phases.

Nonetheless, limnologists easily talk about water chemism as if it were part of God's creation. They seem to forget that the chemical parameters are given by the petrologic composition of the hydrographic basin. Through physical and chemical erosion, the elements are taken off, carried and deposited into the lacustrine basin. In this case, the lake becomes the depositary of the elements within the basin of influence—the hydrographic basin, in our case.

The classifications based upon delimiting the wetlands depending on a set of parameters had satisfactory outcomes. From this perspective, the American school delimits the wetlands depending on hydrologic, geomorphologic, pedologic, and biological parameters. The inter-conditioning of the four parameters leaves no room for ambiguity. Nonetheless, the difficulty of delimitation, the long duration of the research, and the high-analysis costs are still an issue.

The relationship between wetlands and lacustrine basins is extremely close. This is why lacustrine waters are often mistaken for a typical wetland. Most morpho-hydrographic forms, such as wetlands, are only elements of lacustrine basins or stagnant waters. The most important problem is related to the interdependence between morpho-bathymetry and the limno-ecological parameters within various waters.

A mountain unit is mostly known for its altimetric parameters. Each altimetric layer determines a certain type of climate. In their turn, the climatic parameters determine the distribution of plants and animals. The lacustrine basin is the reversed image of a mountain unit. This time, each bathymetric stage corresponds to a cumulus of hydrometric factors. In this case, these are bathymetric layers. Morpho-bathymetry is the edifice on which secondary components are set, just like a house waiting for its dwellers.

The natural dam lakes in Romania have been studied mainly by geographers and less by biologists. Most studies are dedicated to the way they were formed and to emphasizing the evolution of the lake basin in time and space. Most natural dam lakes in Romania are situated in the Eastern Carpathians, where it rains often and the geologic substrate allows landfalls. Most landfalls are favored by the massive deforestations that occurred at the end of the nineteenth century and the beginning of the twentieth century.

The limno-geographic literature until 2012 demonstrated that the Red Lake is the largest natural dam lake in Romania. In 2012, topographic measurements were taken for the most important natural dam lakes and the largest one is Lacul Crucii, measuring 12.95 ha, compared to the 12.01 ha of the Red Lake. As far as the depths are concerned, the situation is the same: 16.3 m for Lacul Crucii and 10.5 m for the Red Lake.

Among the large natural dam lakes in Romania, the Red Lake seems to be the oldest, as it was formed in 1837. Lacul Crucii is the youngest such lake, with two formation stages: in 1978, when a small lake appeared and in 1991, when a big landslide occurred and the landslide body totally blocked the Cuejdul creek.

Because of its relative isolation, the Red Lake caught rather late the attention of interdisciplinary research. The construction of the connection road between Transylvania and Moldavia, over the Eastern Carpathians, meant the end of the lake's isolation. The inclusion of the Red Lake within the transit tourist circuit stimulated the first limno-geographic research. Unfortunately, most studies focused only on the lake basin and on the geologic substrate and less on the physical–chemical parameters and on the water dynamics. The most recent studies focus on the exploitation of the tourist side of the lake and of the surrounding area at the same time. This has brought an invigoration of local tourism, which had suffered a significant decline after 1989, following the Romanian Revolution. The Red Lake exerts a special tourist attraction because the local landscape is ideal for spa-related and climatologic exploitations if community funds can be obtained.

The name of the Red Lake is related to the existence of the reddish limestones on the left bank that often reflect, at sunrise, in the mirror of the lake. The images are also accentuated by the existence of fossil trees bent over and broken from the lake basin.

A chronicle of the second half of the nineteenth century mentions the formation of the Red Lake (Mihaly and Laszlo 2003) as follows: "One night of July 1837, around seven o'clock, from east massive clouds gathered and came rapidly towards our mountains, accompanied by quick lightning, dreadful thunders and, immediately, a heavy rain darkened the entire horizon. The lightnings would not stop, as though foretelling a terrible apocalypse with thunders and endless rains.... People and animals were equally frightened because the lightnings set fire to several households in Gheorgheni and in the surroundings, and the horrible storm went on powerfully until dawn. This caused two landfalls in the Bicaz County, reason for which the Bicaz creek was dammed and it created the Red Lake where, even today, one can find the most beautiful trout".

The origin of this lake also created a very interesting legend related to the interpretation by the local population. In the land of Giurgeu once lived a girl of rare beauty, known by the name of Estera. Her hair was like the feathers of ravens, her eyes were green as olives, and her figure resembled a wind-caressed pine tree.

Estera was extremely hardworking and she was so handy that all the other youngsters asked her favors. One sunny July morning, Estera went to the fair of Gheorgheni. In the market, she met a tall, handsome young man, so strong that he could kill the biggest bear in the area by one stroke. He could also flute like no one else in the land and he was known to have the handiest skills. It was definitely love at first sight, as love comes unexpectedly and it only conquers the hearts of young people.

The young man brought Estera a sky-blue headscarf and he asked her to be his loved one. The two lovers could not marry because the young man was enlisted. However, Estera waited for her man impatiently. In the evening, when the proud sun began hiding beneath the tall mountains, Estera took her trug and went to the smooth source at the foot of the mountain, where she would remain for hours, thinking of her beloved. Even the surrounding mountains softened at her sighs and blue songs.

One Sunday morning, an outlaw saw Estera and—conquered by her beauty—he kidnapped the young woman and ran with her up to the stream called Suhardul Mic, between the thousand-faced rocks, where his entire gang dwelled. The outlaw promised Estera gold, silver, and many other treasures, would she become his wife. Nonetheless, Estera rejected him because she wanted to be faithful and wait for her lover.

The outlaw's perseverance made Estera ask the mountains for help. The rocks heard her desperate prayer. One night in July, a very powerful storm came out of nowhere, dragging along a heavy rain, with horrific thunders and blinding lightning. Making a terrible noise, the mountain fell over and buried the young woman, the outlaw, and also a shepherd and his entire herd.

On the morning of the last Sunday in July, the sun slightly caressed the rocks. The valley, where a day before the clear Vereşcheu creek murmured, was blocked by the landfall, thus creating the Red Lake. If you look in the lake's waters, you can see the red tears and the brightness of the teary, green eyes of Estera (Mihaly and Laszlo 2003). This book tries to underline the role of morpho-bathymetry in the distribution of areas occupied by lacustrine wetlands. At the same time, the physical–chemical characteristics and the water dynamics are underlined, in direct connection with other physical and geographic factors: geologic, geomorphologic, climatic, hydrologic, biologic, anthropic, etc.

In this particular analysis, the Red Lake was chosen because, besides St. Anne's (volcanic) Lake and Lake Bâlea (glacial), it is one of the most important tourist sites in Eastern Romania.

Reference

Mihaly S, Laszlo D (2003) Lacul Roşu şi Cheile Bicazului. Ghid turistic ilustrat. Mark HousePublishing House, Gheorgheni

Apavaloaie M (1971) Contribuţii la cunoaşterea regimului temperaturii aerului din zona Lacului Roşu-Cheile Bicazului. Lucrările Staţiunii de Cercetări Stejaru: 227–237

Chapter 1
Definition of Lakes and Their Position in the Romanian Territory

Abstract Limnology is one of the most important disciplines of hydrology. It studies inland waters and all other standing waters (natural or artificial), including hydrologic phenomena, physical or chemical, in relation to the environment (chemism, thermals, dynamics, development capabilities of the flora and fauna associations, etc.), as well as the way they are valorized by man. From a historical point of view, limnology is only approximately one century old and the founder of limnology is considered to be François A. Forel from the University of Lausanne (Switzerland). It is very important to make a clear distinction between similar terms, such as "lake" and "pool" and define a clear terminology. Out of the grand total of 1 million lakes, distributed all around the world, in Romania, there are currently about 3,450 lakes. There are many types of lakes and numerous classifications are based on several factors, such as: the origin of the lacustrine basin, hydric regime, thermal regime, mineralization degree, trophic potential, geographic position, nature (natural vs. anthropic). The anthropic category is devised into smaller types, depending on the size (the amount of the water reservoir), or on the purpose (hydro-energetic, drinking or industrial water supply, irrigations, pisciculture, recreation, balneotherapeutic, residue cleaning, wet concentration, etc.). The Red Lake is a natural dam lake, which was formed after a huge landslide blocked the Bicaz Valley in 1837. Here, the objective is emblematic, because it is the most well-known natural barrage lake in Romania.

1.1 Concept and Definition

Limnology is the science that studies inland waters and all other standing waters (natural or artificial), including hydrologic phenomena, physical or chemical, in relation to the environment (chemism, thermals, dynamics, development capabilities of the flora and fauna associations, etc.), as well as the way they are valorized by man.

G. Romanescu et al., *Limnology of the Red Lake, Romania*,
DOI: 10.1007/978-94-007-6757-7_1,

The word "limnology" comes from the Greek *limne* = lake. In the category of lacustrine waters are included lakes, ponds and mires. More recently, wetlands have also been included in this category.

Through its subject, limnology is a border discipline between geography and biology. It belongs to geography because it studies depressions and the way they were formed; it studies water, with all its characteristics (physical, chemical and dynamics), meaning the biotype—this sub-branch is called physical limnology. It also belongs to biology because it studies the flora and fauna of stagnant waters (biohydrocenosis)—this branch is called biological limnology. As the biotope and biohydrocenosis within a lake form a whole, standing waters in general represent the most typical systems in nature; they may also be called limno-systems (Hutchinson 1957; Dussart 1966, 1992; Gâştescu 1979).

"Lakes, moreover, form more or less closed systems, so that they provide a series of varying possible ecologic worlds which permit a truly comparative approach to the mechanisms of nature" (Hutchinson 1957). The American biologist Forbes (1887) compares the lake with a microcosm—hence the importance of lake studies to reveal the geographic evolution of an area.

In a broader perspective, limnology is "the interdisciplinary science which studies water basins with slow water exchange—natural or artificial—historically, meaning dynamically in time and regionally in space, the current physical and biotic process, on the grounds of energy and matter transformation within the water body. It also studies the laws of their evolution and geographic status, their natural resources, and possibility of use by mankind" (Gâştescu 1971).

1.2 Limnological History

The Swiss naturalist and professor of medicine, François A. Forel, at the University of Lausanne (Switzerland) (1841–1912) is considered to be the founder of limnology. His first study was "Matériaux pour servir à l'étude de la faune profonde du lac Léman" (1869). The founding book of limnology is "Le Léman: Monographie limnologique," Lausanne, (1892, 1895, 1904) (3 volumes).

Murray (1900) introduces a methodology specific to oceanography and hydrology in his exhaustive research regarding Scottish lakes. Halbfass (1903, 1923) published his treatise on the geography, physics and chemistry of lakes (Berlin), a true basis for modern limnology. Thienemann (1925, 1926) is the first to have studied the insects of streams; he is the most important representative of European limnology (Arlinghaus et al. 2008).

For better organization, on 1st January, 1922, in Kiel (Germany), Thieneman August and Nauman Einar founded the International Society of Limnology (S.I.L.—Societas Internationalis Limnologiae). It comprised 103 specialists from various countries.

The first systematic observations on the lakes in our country truly appeared after 1950. For this organized start, the following geographers should be mentioned: Vîlsan G., Brătescu C., Martonne Emm. de., Morariu T., Coteţ P., Mihăilescu V., etc. There have also been geologists, such as Munteanu-Murgoci Gh., Mrazec L., Maxim I., etc.; biologists: Antipa Gr., Borcea I., Lepşi I., Antonescu C.S., etc.; chemists: Poni P., Petreanu P., etc.

Among the most important geographic limnology works, the following papers, genuine regional monographs must be cited: "Câteva aspecte privind bilanţul hidrologic al lacurilor din Câmpia Moldovei" (Schram 1968); "Lacurile dulci din Câmpia Transilvaniei" (Săndulache 1970); "Lacurile din România" (Gâştescu 1971); "Lacurile glaciare din Carpaţii Meridionali" (Pişota 1971); "Lacurile de pe litoralul românesc al Mării Negre" (Breier 1976), etc.

1.3 Definition of the Terms Lake and Pool

Lake is "a body of standing water occupying a basin and lacking continuity with the sea" (Forel 1892). The term lake applies to all natural or artificial depressions containing water, "with their own life and certain autonomy" (Dussart 1966, 1992).

One of the simplest definitions of the term lake says that it is "an inland body of water of considerable size" (International Glossary of Hydrology 1992). At the same time, lake is "a sweet or salt water body which occupies a closed depression on the surface of a continent" (Dictionnaire Encyclopédique Alpha 1983).

In defining the notion of lake, the main criteria should be the basin and water body—which are inseparable—on the one side and the hydric characteristic of the relatively stagnant water, on the other. Consequently, "a lake is a relatively stagnant water volume, cantoned in an inland depression, with no continuity with the Planetary Ocean" (Gâştescu 1971). According to this definition, the natural and artificial basins also fall into the category of lakes (Bratsk, Assouan, The Iron Gates I and II, etc.), as well as the water bodies of thousands of square meters and thousands of square kilometers (Caspian Sea, Aral, Baikal, Victoria, etc.). In the category of lakes the authors also included the water bodies above or below the Planetary Ocean level (Dead Sea, Lake Tiberias, etc.), at a depth of several dozen meters or thousands of meters (Baikal, Tanganyika, Malawi, etc.). They are called lakes regardless of their emersed or submerged vegetation, productive or unproductive bottom, whether they are salty (Dead Sea, Caspian Sea, Techirghiol, Ursu, etc.), brackish (the Razim-Sinoe lagoon complex) or sweet, if they are permanent or temporary, etc. For this definition, the authors are interested in the form of the depressions where the water is cantoned (circular, oval, elongated, digital, etc.). The main characteristic is that of the slow water exchange with the environment; it triggers a series of particularities, of the hydric, thermal and hydrochemical regime, of the lake-specific hydrocenoses, etc.

Of the various definitions so far, there is not a clear-cut distinction between lake, pond and pool. The transition of these lacustrine categories is not entirely transparent. The size cannot be the only main criterion, as a pond can be much larger than a small lake. Depth has also been used as a distinction index, which, through the physical-geographic indices, allows a definition of the classification limits.

The consensus is that (Imboden 1976) a lake presents, besides the littoral area, a deep area with no light and where the bottom is covered by vegetation. Its water is often stagnant, as the currents are present mostly when the water stream has tributaries or important emissaries (Fig. 1.1). Certain Romanian authors have also accepted this definition, such as Antonescu (1967). He states: "We call lake a standing water, so flat and deep, that we have two distinct areas: (1) a bank (littoral) area covered by aquatic emersed and submerged vegetation, hence with a productive bottom, and (2) a deep (profound) area, obscure, with no plants and covered, on the unproductive bottom, with a fine, uniformly structured silt." "According to this limnological definition, Romania has only a few lakes, given the above mentioned meaning: Snagov (11 m deep) and certain standing waters in the Carpathians (Zănoaga, Bucura, and Galeş—the first 27 m deep), the Red Lake and the Big Bicaz Lake (around 93 m). Nonetheless, Romania has numerous standing water surfaces, some very large but not at all deep—this is why the vegetation can grow anywhere on their bottom. These waters are called ponds, or sometimes tarns and mires."

With regards to ponds, they are never deeper than 3 m and they can have their entire bottom covered with aquatic plants that do not need sunlight. Standing waters, because of their small sizes and volume, are sensitive to the variations in external influences. If the accumulations are artificial, a device is needed to make sure there is always a level-up. In Romanian geographic literature, a pond is, most

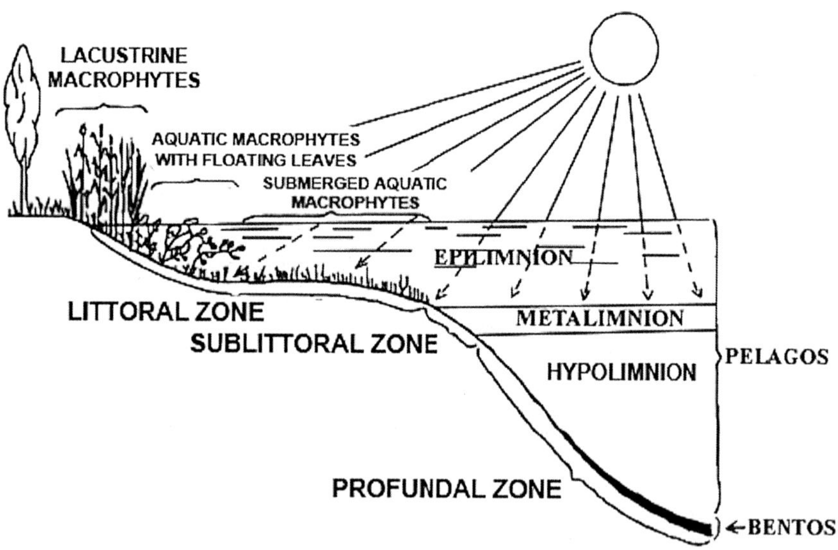

Fig. 1.1 Zoning of the lacustrine environment

of the times, an artificial lake. The pool of water is a small inland water body, usually not deeper than 80 cm, where the sun affects all the water layers; in this case, there are daily variations in the temperature. The smallest water surfaces are called puddles; during the dry season most of them disappear. In the case of our country, the term pool should be used in its broader, popular meaning only for the water surfaces within floodplains. In this case, it is not about a certain lake but about a geographic area, a natural unit with several elements: geographic units, hydrography, i.e., vegetation, fauna, economic use, etc. It may have two meanings: a toponym specific to floodplains (the Danube or Siret Floodplain, etc.); or a toponym limited to plain lakes of genetic type but not as an evolution phase (it is not widely known for this meaning). de Emm (1902) mentions that "At Călăraşi begins what people name the Pool."

"The Romanian people (the ones near the Danube, and also the shepherds from Ardeal who have taken their sheep there for centuries) name the Danube Floodplain the Pool, meaning the natural geographic area within the country… a natural unit with a similar role to that of the Carpathians in the history of our people" (Conea 1965).

From the definitions given to lakes and to standing waters in general several conclusions have resulted. The most important feature is the stagnant character of the water, which is also relative. This is the main difference compared to streams.

Dimensions (surface and depth) represent the second feature. From this perspective, the definitions of Imboden and Antonescu are based on biological criteria (Gâştescu 1998). In this case, the Sea of Azov, with only 9 m depth and without a profound area, should be included in the category of ponds or pools.

The third feature is that of the lake position compared to a marine basin. If the definitions imply that the lake has nothing to do with the sea, lagoons should also be eliminated. Most of the times, lagoons may be classified within transition systems or they may be clearly delimitated, in which case they are included in the category of lakes. Their features have been strongly altered by anthropic activities (the level of the Razim-Sinoe lagoon complex is 50 cm higher than that of the Black Sea, the water salinity is down by 2–5 ‰, etc.). According to the last feature, the Caspian Sea is truly a lake.

1.4 Distribution of Lakes

There are around 1 million lakes in the world, which represent between 2.1 and 2.7 million km^2 (1.4–1.8 % of the land, meaning the size of the Mediterranean Sea). The total volume is approximately 700,000 km^3 (Pisota and Nastase 1957; Pişota 1995; Zăvoianu 1999).

In Romania there are around 3,450 lakes (of which 2,300 are natural and 1,150 are artificial), with a 2,620 km^2 surface (Gâştescu 1971).

- The lakes within areas with a humid and temperate climate

They have large volumes of water and they are the most numerous, with a pluvial, pluvio-nival, nivo-pluvial or nival supply regime. Most of them become sweet and they present surface outflows. They form the category of outflowing lakes because they are connected to the Planetary Ocean (e.g., Onega, Ladoga, Baikal, Biwa, Tanganyika, Victoria, Huron, Ontario, Titicaca, Balaton, Zănoaga, etc.).

- The lakes within arid and dry areas

They are not numerous and they present no outflows. They have a small water volume; most of the time they are salty, with a pluvial or pluvio-nival supply regime. They are not connected to the Planetary Ocean and they form the category of non-outflowing lakes.

In Africa they are called chotts (salt water lakes or big *sebkha*); *sebkhaouri* (*sebka*) (arid saline depressions found in northern Africa and Arabia; geologically speaking, they represent deposits of gypsum and anhydrite mixed with sand and silt); and salt pans (ex: Ciad, el Hodna, Chott el Jerid, the Dead Sea, Aral, the Caspian Sea, etc.).

1.5 Types of Lakes

Since the first hydrologic books in general and limnological in particular, lakes have been categorized depending on their dominant characteristics.

The best-known classifications are the following:

(a) depending on the origin of the lacustrine basin: resulting from the action of internal factors (tectonic, volcanic); resulting from the action of external factors (from streams; from water dissolution; glacial; of wind depressions; of sea and ocean banks, etc.);
(b) depending on the hydric regime: permanent; temporary; with outflow; without outflow (closed);
(c) depending on the thermal regime: polar; temperate; tropical; cold; mesothermal; thermal;
(d) depending on the mineralization degree and on the chemical composition: sweet; brackish; salty; hydro-carbonated; sulphated; chlorinated;
(e) depending on the trophic potential: oligotrophic; eutrophic;
(f) depending on the geographic position within various landforms: mountainous; plateau; plain; littoral, etc.;
(g) depending on their nature: natural; anthropic (artificial).

For the artificial (anthropic) lakes, the following characteristics are taken into account as classification criteria:

- the purpose (hydro-energetic, drinking or industrial water supply, irrigations, pisciculture, recreation, balneotherapeutic, residue cleaning, wet concentration, etc.);
- the size (the amount of the water reservoir).

Depending on the origin of the lacustrine basin

This classification is specific to geography and geology. Most classifications of this type have as grounds the works of Penck (1882), von Richtofen (1886), Davis (1887), Russell (1895), Delebecque (1898), Geikie (1905), de Emm (1909), Collet (1925), Mehedinţi (1930), Hutchinson (1957), Dussart (1966, 1992), Gâstescu (1971), Guilcher (1979), Pişota and Buta (1983), etc.

At the beginning, the geomorphologists took into account the constructive, destructive and obstructive factors. Another element to be taken into consideration is time (Davis 1887). Consequently, a lake cannot be permanent; it appears, it develops and then it disappears (Wetzel 2001).

Hutchinson (1957) made the most complete classification by the origin of the lacustrine basin. The author determines the existence of 76 genetic types of lacustrine basins, classified into 11 groups of natural processes:

I. Tectonic Basins

1. Relict lakes isolated from the sea by epirogenetic movements;
2. Gentle epirogenetic uplift of irregular marine surfaces;
3. Reversal of the hydrographic pattern by tilting or folding;
4. Large-scale basins formed by warping;
5. Earth movements producing local subsidence;
6. Lakes in basins in tectonically dammed synclines;
7. Old peneplain surfaces as intermontane basins;
8. Basins associated with fault scarps in a tilted terrain (half-graben);
9. Basins associated with fault scarps in an elongated depression (graben or series of half-grabens).

II. Volcanic Lakes

10. Unmodified craters in a cinder cone;
11. Explosion craters;
12. Maars (abortive embryonic volcanic depression);
13. Calderas;
14. Conche (series of craters producing terraces);
15. Modification of caldera lakes by secondary activity;
16. Volcano-tectonic basins;
17. Lakes on collapsed or irregular lava flows;
18. Lakes formed through volcanic damming by peaks or secondary peaks;
19. Lakes formed through volcanic damming by a lava stream or mudflow.

III. Landslide Lakes

20. Lakes formed through dams formed by rock flows, mud flows and debris flows (Lake Quake-California, Clear-California);
21. Lakes formed between the mass of a slide and a detached valley wall (Red Lake, Lake Crucii - Romania; Lake Brazeau-Alberta (Cruden 1982), Sarez-Pamir, Chaillexon-Franţa, Busyû-Japonia) (Gâştescu 1979);
22. Lakes formed by the building of scree dams across valleys (Lake Arrowhead-California).

IV. Glacial Lakes

23. Lakes on or in ice;
24. Glacier-dammed lakes;
25. Damming by moraines of existing glaciers;
26. Ice-scour lakes;
27. Cirque lakes;
28. Paternoster lakes, fiords and piedmont lakes (valley rock basins);
29. Glint lakes (ice cauldrons);
30. Damming by terminal or recessional moraines (no glacier);
31. Damming by outwash in a valley;
32. Lateral valley dammed by a terminal or lateral moraine in the main valley;
33. Lakes between terminal deposits;
34. Lakes formed in irregularities in ground moraine;
35. Kettle lakes following the valley drainage pattern from an ice block with outwash;
36. Kettle lakes from stagnant ice blocks within outwash (no pattern);
37. Kettle-hole lakes (ice blocks within an outwash in an unglaciated valley);
38. Kettle lakes in till;
39. Kettle lakes associated with esker development;
40. Glacial tunnel lakes;
41. Lakes formed due to local thawing of permafrost;
42. Lakes formed by melting of ice wedges on permafrost.

V. Solution Lakes

43. Karstic depressions (dolines (funnel), uvala (compound funnel));
44. Poljes (tectono-karstic depression);
45. Underground lakes in caverns;
46. Solution lakes in gypsum;
47. Solution lakes in ferric hydroxide and hydrous aluminum silicate.

VI. Fluviatile Lakes (From Running Water)

48. Plung pools and evorsion lakes;

49. Lakes formed by fluviatile dams;
50. Fluviatile damming within lakes;
51. Strath lakes (parallel sediment ridges formed in a long narrow basin);
52. Lateral lakes (lateral valleys dammed by deposits of a main stream);
53. Deltaic levee lakes;
54. Meres (levee lakes influenced by tidal forces);
55. Oxbow lakes;
56. Lakes formed by uneven aggradation in a floodplain;
57. Lakes formed by the abandonment of well-defined channels;
58. Lakes formed between a scarp (edge of floodplain) and a levee;
59. Crescentic levee lakes.

VII. Wind Lakes

60. Lakes formed by damming through sand dunes;
61. Lakes formed through uneven deposition of loess;
62. Interdunal lakes;
63. Deflation basin lakes.

VIII. Coastal Lakes

64. Lagoonal or coastal lake (river discharge vs. constant tide level);
65. Lakes formed when two tombolos connect an island to the coast;
66. Lakes formed along a lake shore when a bar is built across a bay;
67. Lakes formed when a lake is bisected by two spits;
68. Lakes formed behind a cuspate spit of a larger lake.

IX. Organic Accumulation Lakes

69. Phytogenic dam lakes (peat bogs);
70. Closed phytogenic basin lakes (peat bogs);
71. Lakes in coral atolls.

X. Mammal-Constructed Lakes

72. Beaver dammed lakes;
73. Reservoirs;
74. Lakes in artificial depression.

XI. Meteoritic Impact Lakes

75. Lakes formed in craters formed by meteoritic impact;
76. Bay lakes formed through compression waves of impact.

Throughout time, the classification made by Hutchinson in 1957 has been completed or simplified depending on the factors considered. Most classifications after 1957 had this classification as a reference point. The Romanian literature borrowed integrally the directives of English literature (Romanescu 2003).

Of the entire lake surface, the glacial and periglacial lakes (1,247,000 km^2) were underlined, followed by tectonic lakes (524,000 km^2). As for the volume, the first are the tectonic lakes (56,600 km^3) and the second the glacial and periglacial lakes (38,400 km^3; Cohen 2003) (Table 1.1). As far as volume is concerned, Lake Baikal is the largest, with 23,615 km^3, which represents 20 % of the global freshwater volume.

The most usual classification in Romania targets the geographic position of a lacustrine basin within the main landforms. The primordial elements concern the altitude and the thermal gradient. Depending on this criterion, the lakes are classified into the following groups:

- mountain lakes: Red (Roşu) (Fig. 1.2), Bolătău, Ştiol (of a natural dam), Bucura, Zănoaga, Lala, Buhăescu, Bâlea (glacial), St. Ana (volcanic), Ocna Şugatag, Telega, Vintileasca (clasto-karstic), Izvorul Muntelui, Vidraru, Vidra, Călineşti-Oaş, Firiza, Bodi-Mogoşa, Colibiţa, Bâtca Doamnei, Mesteacănu, Frumoasa, Poiana Uzului, Săcele, Brădişor, Siriu, Măneciu, Paltinu (anthropic) (Fig. 1.3), etc.;
- hill and plateau lakes: Ursu, Cojocna (clasto-karstic), Crucii (natural dam), Balta Vulturului, Balta Roşie (of subsidence), Dracşani, Podul Iloaiei (ponds), etc.;
- plain lakes: Sărat, Amara, Lătenilor (of meandre); Movila Miresii, Ianca, Strachina (sinkholes), Băneasa, Herăstrău, Mogoşoaia (anthropic), etc.;
- littoral lakes: Razim, Sinoie, Siutghiol (lagoons), Mangalia, Tatlageac, Agigea, Techirghiol (fluvio-marine limans);
- floodplain lakes and the Danube Delta: Bugeac, Oltina, Mârleanu (fluvial limans), Crapina, Brateş (floodplain lakes), Sireasa, Dranov, Gorgova, Furtuna (delta lakes), Erenciuc, Belciug (of meander) etc.

The Red Lake is part of the water bodies formed after damming streams to prevent landslides. Landslides may be caused by heavy rains or by earthquakes. In our case, the issue is still undetermined, even though rain is more likely the cause.

On the Romanian territory there are lakes formed because of landslides only in the Eastern Carpathians and Subcarpathians: Crucii, Ştiol, Vulturilor, Bolătău, Iezer, Dracului (Romanescu 2003; Mîndrescu et al. 2010). It appears that the oldest natural dam lakes in Romania are over 400 years old (Bolătău, Iezer). They are situated in the Eastern Carpathians and they were caused by landfalls (Mîndrescu et al. 2010).

Table 1.1 Aggregate areas and volumes of the major lake classes at the global scale [After Meybeck (1995) and Herdendorf (1990)] (Cohen 2003)

No	Lake type	Aggregate area (km^2)	Aggregate volume (km^3)
1	Glacial and periglacial lakes	1,247,000	38,400
2	Tectonic lakes[a]	524,000 (893,000)	56,600 (134,900)
3	Fluvial lakes	218,000	580
4	Coastal lakes	40,000	130
5	Volcanic lakes	3,150[b]	580[b]
6	Other lake types	88,000	300

[a] Values including the Caspian Sea are shown in brackets

[b] Crater lakes only

Fig. 1.2 The Red Lake and the Suhard Mountain

Fig. 1.3 Geographic position of the main lakes in the East Carpathian Mountains

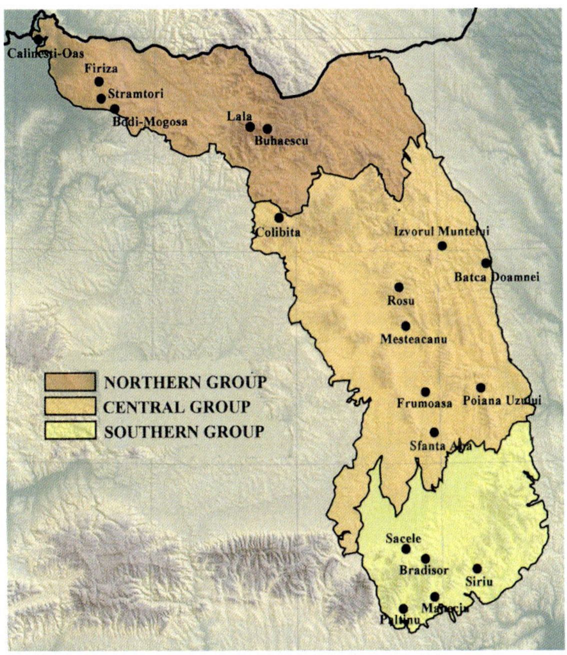

1.6 Natural Dam Lakes and Their Spatial Distribution

Several conditions have to be met for the formation of natural dam lakes: packs of permeable rocks situated above impermeable rocks (most of the times clays); elevated slope; high discharge, usually after long-term heavy rains; tight valley, with abrupt walls; lateral erosion, in the versant, exerted by streams; high-discharge river to supply a newly formed lacustrine basin, etc. (Costa and Schuster 1988; Glade and Crozier 1999; Meyer and Leidecker 1999; Neuffer and Bruhn 2005; Noord and Phuntsho 2005; Schneider et al. 2004). The rock packs may also be formed after excessive precipitation or earthquakes (84 %). Volcanic eruption accounts for a further 7 % of dams. Other causes of landslides account for the remaining 9 % (Schuster and Costa 1986).

The most important cause, indirect, is the massive deforestation that has occurred mainly in poor or developing countries. The disappearance of forests also eliminates the diluvium support through the root system. The accelerated pluvial denudation phenomenon washes the slopes and makes the water infiltrate into the crevice sectors formed during the hot or dry season (Schneider et al. 2011).

On the Romanian territory there are such favorable conditions in the Eastern Carpathians, Moldavian and Curvature Subcarpathians, as well as in the Moldavian Plateau. The most favorable conditions are specific to wildflysch areas, with sequences of permeable and impermeable sedimentary layers, situated in the middle sector of the Eastern Carpathians.

In the world, there are such conditions in the mountain ranges of Himalaya, the Andes, the Rocky Mountains, Pamir, Alpes, Tian Shan, etc. (Huss et al. 2007). There are conditions for landslides, most of the times on the surface, in subarctic areas, with winter and summer thermal alternations. In the few months of heat, the active layer slips down the permafrost and creates landslide masses that block the water streams (Marsh and Neumann 2003). Because the landslide diluvium is small, the lake is temporary. Such cases occur in Alaska, Labrador, Taymyr, Laponia, Land of Fire, etc. (Romanescu 2005).

Landslides—landslide dams—that move an important amount of material usually determine the dams blocking the water streams. They can also be formed by the materials carried by the rivers during floods, after rock falling, as a consequence of the volcanic activity, eolian processes (dune-dammed), or they may be carried and transported by various animals, as for example by beavers (debris dam). The generic name is dam lakes or natural dams.

"Of the numerous kinds of dams that form by natural processes, dams formed from landslides, glacial ice, and late-neoglacial moraines present the greatest threat to people and property. The most common types of mass movements that form landslide dams are rock and debris avalanches; rock and soil slumps and slides; and mud, debris, and earth flows. The most common initiation mechanisms for dam-forming landslides are excessive rainfall and snowmelt and earthquakes" (Costa and Schuster 1988).

Dams, which can enclose a lake surface, can be caused by a set of factors. Besides the dominant factors (such as landfalls and glaciations), those related to volcanism fluvial, wind-related or organic processes (Table 1.2) must be mentioned. The most dangerous dams—which can be broken and can cause material damage or loss of human lives—appear after landfalls, because of mire or rock accumulations, out of ice masses or behind moraines. Some natural dams have economic benefits, such as hydropower generation and recreation (Anderson 1948; Adams 1981; Jones et al. 1985; Loope et al. 1995).

The most frequent—and the largest and most diversely shaped at the same time—natural dams are a consequence of landslides. At the same time, when there is a dam breaching, a catastrophic flood can occur. In this sense are mentioned

Table 1.2 Types of obstructive natural dams [Costa and Schuster (1988) modified from Davis (1882) and Hutchinson (1957)]

No.	Type of dam	Example
Volcanic dams		
1	Volcanic peaks	Lake Nicaragua, Nicaragua (Hutchinson 1957)
2	Lava flows	Snag Lake, California (Finch 1937)
3	Pyroclastic flows	Rio Magdalena Lake, Mexico (Silva et al. 1982)
Landslide dams		
4	Slides/slumps	Earthquake Lake, Montana (Hadley 1964)
5	Mud/debris/earth flows	Lake San Cristobal, Colorado (Crandell and Varnes 1961)
6	Rock/debris avalanches	Spirit Lake, Washington (Meyer et al. 1986)
7	Liquefaction of sensitive clays	Yamaska River Lake, Quebec, Canada (Clark 1947)
8	Peat slides	Addergoole Bog Lake, Ireland (Ousley 1788)
9	Scree	Goatswater, United Kingdom (Marr 1916)
Glacial dams		
10	Ice	Gapshan Lake (Shyok), Pakistan (Mason 1929)
11	Moraine	Nostetuko Lake, British Columbia, Canada (Blown and Church 1985)
12	Ice and snow avalanche	Rio Plomo, Argentina (King 1934)
Fluviatile dams		
13	Tributary sediments	Lake Pepin, Minnesota-Wisconsin (Davis 1882)
14	Main-channel sediments (lateral lakes)	Lake Tung-ting, China (Hutchinson 1957)
15	Alluvial fans	Lake Tulare, California (Hutchinson 1957)
16	Deltas	Blue Lakes, California (Davis 1933)
17	Levee deposits (oxbow lakes)	Old River, Louisiana (Campti quadrangle, U.S. Geological Survey)
Eolian dams		
18	Dunes	Moses Lake, Washington (Russell 1893)
Coastal dams		
19	Bay-bars	Freshwater Lagoon, Eureka, California (Cotton 1941)
Organic dams		
20	Logs and other vegetation	Lake Okeechobee, Florida (Hutchinson 1957)
21	Beaver dams	Beaver Lake, Montana (Hutchinson 1957)

Switzerland in A.D. 563 (Eisbacher and Clague 1984) and central Java in A.D. 1006 (Holmes 1965). As regards the floods that have caused heavy loss of human lives, the Deixi dambreak in 1933 on the Min River in central China should be mentioned, when 2,423 people died (Li et al. 1986); the dambreak on the Indus River in 1841 (Mason 1929). The world's worst recorded landslide-dam disaster occurred when the 1786 Kangding-Louding earthquake in Sichuan Province, China, triggered a huge landslide that dammed the Dadu River. After 10 days, the landslide dam was overtopped and breached; the resulting flood extended 1,400 km downstream and drowned about 100,000 people (Schuster and Evans 2011). The peak discharges of lake waters can reach extremely high values (Table 1.3).

Most natural dams created after landslides appear in mountain areas, with narrow valleys and abrupt slopes. They are specific to areas with seismic, volcanic, or glacial activity (Costa and Schuster 1988). Narrow valleys facilitate the accumulation of a relatively low volume of sliding material. In this case, the dam is more enduring. The dams within wide valleys, which intersect the marine of lake deposits, facilitate the accumulation of a large mass of sliding material, which can create huge dams (Clark 1947; Costa and Schuster 1988; Evans 1984).

Landslides, slope falling, or mudslides cause most natural dams. Rarely, natural dams appear after a fall of clayish slopes or rocks. In this case, there is a limited area affected because a small amount of material is set in motion. The big natural dams, formed after landslides, are caused by complex manifestations, such as landfalls, mudslides, or slope falling. The most eloquent example is the accumulation of 2.8 km^3 of rocky material and of mud sliding material causing the biggest landslide in history, produced during the eruption of Mount St. Helens in 1980 (Costa and Schuster 1988). On the valley of the North Fork Toutle River, five large lakes emerged; only three still exist today (Meyer et al. 1986).

The natural dams formed after mudslides or by accumulations of excessively carried alluvia are not high and their deposit is non-cohesive. In this case, they only last for a couple of weeks, months, or years. When dams break, there are no catastrophic floods because the water volume accumulated in the lake is rather low (Li et al. 1986). Only the cohesive volcanic rocks can create more long-lasting dams [the dam on the Gunnison River in Colorado, which created Lake Fork around 700 years ago (Crandell and Varnes 1961)]. The most typical dams that emerged after the fall of rock blocks were those of 1943 on Grande Riviere du Chene in Quebec (Canada) and of 1966 in Russia (Lake Yashinkul) (Pusharenko 1982) (Table 1.4).

The most important causes contributing to landslides are represented by precipitations (rains and snow melting) and earthquakes (90 % of the cases) (Costa and Schuster 1988; Romanescu and Jigău 1998). Another frequent cause is volcanism (8 %). Most natural dams form after a single landslide or earthquake. On the Totsu River in Japan, 53 landslides occurred on a surface of 1,000 sq km in 1889 (Swanson et al. 1986). In 1929, in South Island, New Zealand, an earthquake with a magnitude of 7.6° took place and it triggered a landslide that formed eleven natural dam lakes (Adams 1981). The highest number of natural dam lakes was

Table 1.3 Large ice-dammed lake outburst floods in historic and pre-historic times (Herget 2012)

Location	Date	Lake volume km^3	Peak discharge m^3/s	Reference
Missoula, Montana U.S.A.	Pleistocene	2,184	17,000,000	O'Connor and Baker (1992)
Missoula, Montana U.S.A.	Pleistocene	2,184	10,000,000	Baker and Costa (1987)
Nedre Glåmsjø, Norway	Pleistocene	99	170,000	Berthling and Solide (1999)
Lake Alsek, Canada	Holocene	30	470,000	After Clarke from Clague and Evans (1994)
Kjölur, Iceland	9500 BP	25	200,000	Tómasson (2002)
Hubbart Glacier, Alaska U.S.A.	1986	5,3	104,500	Mayo (1987); Seitz et al. (1986); Mayo (1989)
Lake Alsek, Canada	ca. 1850	4,7	30,000	After Clarke from Clague and Evans (1994)
Lake Elk, Canada	Pleistocene	4	20,000	Clague (1973)
Lake George, Alaska U.S.A.	1958	2,2	10,160	After U.S. Geological Survey from Fahnestock and Bradley (1973)
Lake George, Alaska U.S.A.	1961	1,73	10,050	After U.S. Geological Survey from Fahnestock and Bradley (1973)
Graenalon, Iceland	1935	1,5	5,800	Thorarinsson (1939)
Graenalon, Iceland	1939	1,5	5,000	Thorarinsson (1939)
Lake Batal, Himalaya	Pleistocene	1,496	24,000	Coxon et al. (1996)
Lake George, Alaska U.S.A.	1960	1,48	9,280	After U.S. Geological Survey from Fahnestock and Bradley (1973)
Van Cleve Lake, Alaska U.S.A.	1992	1,4	4,500	After Brabets from Walder and Costa (1996)
Chong Kumdam (Shyo), India	1929	1,35	22,650	After Gunn or Mason et al. from Hewitt (1982)
Lake George, Alaska U.S.A.	1959	1,11	6,310	After U.S. Geological Survey from Fahnestock and Bradley (1973)
Lake George, Alaska U.S.A.	1965	1,11	6,680	After U.S. Geological Survey from Fahnestock and Bradley (1973)

created after the 1783 earthquake in Calabria (Italy), when 215 small aquatic surfaces emerged (Cotecchia 1978).

In the specialized literature, there is also a case cited when several landslides led to the creation and destruction of a natural dam lake; the landslides occurred in different years and various causes led to its destruction (Chang 1984). In 1862, in the Tsao-Ling region in the centre of Taiwan, an earthquake caused a landslide on

Table 1.4 Well-documented examples of landslide dams formed by specific classes of landslides

Landslide class and name	Years	Dammed river	State/country	Landslide volume m^3	Blockage dimensions			Lake dimensions		Dam failed?	References
					Height (m)	Length (m)	Width (m)	Length (km)	Volume (m^3)		
Slides/slumps											
Deixi landslide	1933	Min River	Sichuan	150×10^6	255	400	1,300	17	400×10^6	Yes	Chang (1934); Li et al. (1986)
Lower Gros Ventre landslide	1925	Gros Ventre River	Wyoming, U.S.A.	38×10^6	70	900	−2,400	6.5	80×10^6	Yes	Emerson (1925); Alden (1928)
Tsao-Ling rock slide	1941–1942	Chin-Shui-Chi River	Taiwan	250×10^6 (two slides)	217	1,300	2,000	–	157×10^6	Yes	Chang (1984)
Cerro Condor-Sencca rock slide	1945	Mantaro River	Peru	$5{,}6 \times 10^6$	100	250	580	21	300×10^6	Yes	Snow (1964)
Madison Canyon rock slide	1959	Madison River	Montana, U.S.A.	21×10^6	60–70	500	1,600	10	–	No	Hadley (1964); Knight and Bennett (1960)
Thistle earth slide	1983	Spanish Fork River	Utah, U.S.A.	22×10^6	−60	200	600	5	$78–10^6$	No	Kaliser and Fleming (1986)
Mud/debris/earth flows											
Slumguillon earth flow	1200–1300	Lake Fork, Gunnison River	Colorado, U.S.A.	$50–100 \times 10^6$	40 (rough estimate)	500	1,700	3	–	No	Crandell and Varnes (1961)
Gupis debris flow	1980	Ghizar River	Pakistan	–	30	200	300	5	–	No	Nash et al. (1985)
Polallie Creek debris flow	1980	East Fork, Hood River	Oregon, U.S.A.	$70–100 \times 10^3$	11	–	230	–	$105–10^3$	Yes	Gallino and Pierson (1985)

(continued)

Table 1.4 (continued)

Landslide class and name	Years	Dammed river	State/country	Landslide volume m^3	Blockage dimensions			Lake dimensions		Dam failed?	References
					Height (m)	Length (m)	Width (m)	Length (km)	Volume (m^3)		
Rock/debris avalanches											
Usoy landslide	1911	Murgab River	Tadzhikistan, U.S.S.R.	$2.0–2.5 \times 10^9$	301 (Bolt)	1,000	1,000	53	–	Partial failure	Gasiev(1984)
Tanggudong debris slide/ avalanches	1967	Yalong, China	Sichuan, China	68×10^6	175	650	3,000	53	680×10^6	Yes	Li et al. (1986)
Mayunmarca rock slide/debris avalanche	1974	Mantaro River	Peru	1.6×10^9	170	1,000	3,800	31	670 × 106	Yes	Hutchinson and Kojan (1975); Lee and Duncan (1975)
Mount St. Helens rock slide/debris avalanche	1980	North Fork, Toule River	State of Washington, U.S.A.	2.8×10^9	Avr. = 45	800 (Spirit Lake)	24×10^3 (Spirit Lake)	5.5 (Spirit Lake)	259×10^6 (Spirit Lake)	No (Spirit Lake)	Meyer et al. (1986)
Liquefaction of sensitive clays											
	1898	Riviere Blanche	Quebec, Canada	2.6×10^6	8	400	3,200	–	–	Yes	Dawson (1898)
	1945	Yamaska River	Quebec, Canada	117×10^3	3–4	75	425	–	–	Yes	Clark (1947)

the Chin-Shui-Chi River. The dam disappeared in 1898 because of unknown causes. Another earthquake, of 1941, caused a landslide that formed a 140 m high-natural dam, on the same spot as in 1862. Heavy rain in 1942 reactivated the landslide and the dam became higher, reaching 217 m. The dam broke completely due to heavy rains in 1951, when the flooding caused the destruction of 564 households and the death of 154 people. On the 15th of August 1979, there was another landslide and another natural dam emerged, measuring 90 m in height. Just nine days afterwards it broke, because the rain had not ceased (Costa and Schuster 1988).

From a geomorphologic perspective, natural dams caused by landslides depend on their relation to the valley floor (Swanson et al. 1986).

- Type I—small dams, in contrast to the width of the valley. The dams do not reach from one side to the other.
- Type II—large natural dams that occupy almost the entire valley floor, without reaching from one side to the other.
- Type III—very large natural dams that fill the valley from side to side and have considerable distances upstream and downstream.
- Type IV—natural dams that emerged after the old dams broke and the materials fell on both sides of a valley. New landslides can occur in the central sector of the valley or they can overlap the old ones.
- Type V—natural dams that formed when the same landslide has multiple lobes of debris that extend across a valley floor at two or more locations.
- Type VI—comprises dams created by one or more surface failures that extend under the stream or river valley and emerge on the opposite valley.

After analyzing 184 natural dams caused by landslides, the most common type is II (44 %), followed by type III (41 %), type I (11 %), type VI (3 %) and types IV and V (<1 %) (Swanson et al. 1986; Costa and Schuster 1988).

The most important ones that determined the appearance of lacustrine basins are the following: Usoi Dam (Tajikistan) 301 m; Waikaremona Lake (New Zealand) 250 m; Attabad Lake (Pakistan) 125 m (Kargel et al. 2010); Gros Ventre (USA, Wyoming) 60 m; Quake Lake (USA, Montana) 58 m; Tangjiasham Lake (China, Sichuan) 52 m; Riñihue Lake (Chile) 26 m, etc.

The barriers created on water streams create two types of high- water waves: backwater waves (backflooding, upstream flooding), behind the dam; downstream flooding, after dam breaching. When the level of lake waters rises, underground waters also change their position, with slight increases in the piezometric level. When the level of lake waters decreases, the same happens with underground waters, determining intense landslides, with massive diluvium material.

From a geomorphologic perspective, the creation of a dam determines the valley aggradation upstream; the dam breaching will determine the valley aggradation downstream (Jansen 1988). The type I lakes are small, with low depths and no flood would be registered if dams broke. The type II lakes are larger and more dangerous. As for type III (North Fork Toutle River in Mount St. Helens,

Washington), the lakes are even more dangerous, mostly when they enclose tributaries with adjacent lakes.

The type IV (Da Lake on Min River, China) (Li et al. 1986) and V lakes (Slide Lake in Glacier National Park, Montana) become dangerous if there are narrow valleys and the dam is high (Butler et al. 1986). As for type VI lakes, there is no complete blockage. In this case, the river continues to flow over the dam formed on the central stream of the valley. This is why there is a small water accumulation, which does not imply a real flooding hazard (there are such lakes in Japan, as well as one in Colorado) (Swanson et al. 1986).

As consequences of the above-mentioned factors, the lakes formed behind the landslide waves cannot exist for a long time (Herget and Carling 2004). Historically speaking, they are temporary. The oldest natural dam lakes were formed in 1248, in France (Saint Andrew, on Mountain Garnier) and in 1590, in Japan (Busyû) (Jakobson et al. 2007).

The nature of the landslide material, the piping and pressure of the accumulated water make the dam breach quickly. When the water penetrates the dam, the finest particles, such as clay, are eliminated. In this way, after some while, the dam is formed only of coarser materials, with no binder. If there is a big amount of water accumulated upstream, it falls and facilitates partial or total water elimination. If the dam falls suddenly, floods may occur. If the erosion is gradual, there is only a slight increase in the downstream flow.

"A landslide dam in its natural state differs from a constructed embankment dam in that it is made up of a heterogeneous mass of unconsolidated or poorly consolidated earth material and has no engineered water barrier (impervious zone), filter zones to prevent piping, nor drain zones to control pore pressure. It also has no channelized spillway or other protected outlet; consequently, landslide dams commonly fail by overtopping, followed by breaching from erosion by the overflowing water" (Costa and Schuster 1988).

Most of the times, dams disappear because the river discharge is high and the water is quickly spilled over the crest. In this case, the deep erosion, on the dam, is more than obvious; very soon, a new riverbed is created. This is why most lakes are not noted on maps, or, if they are drawn on old cartographic materials, they are no longer present on newer ones.

The most important dambreaks were caused by the piping of fine-textured soils: the Lower Gros Ventre landslide in northwestern Wyoming, broken in 1927 (Alden 1928); Cerro Condor-Sencca landslide dam in Peru, failed in 1945 (Snow 1964); Lake Yashinkul on the Isfayramsay River in the south-central Rusia (Glazzyrin and Reyzvikh 1968); Cache Creek in northern California (Scott 1970), etc.

The floods after natural dambreaks can be catastrophic. In the specialized literature, numerous such hydrologic events caused by a dambreak in various areas are attested (Costa and Schuster 1988). In order to attenuate the damage caused by dam braking, detailed investigations can be undertaken and probable models for each case can be provided (Dong et al. 2011). In this case, several events are described: the "Spokane flood" in the east of the State of Washington, which occurred around 16,000–12,000 years ago, after an ice dam broke (Missoula of

Table 1.5 Well-documented examples of glacier dams that have failed, producing Jökulhlaups (Costa and Schuster 1988)

Lake name	Location	Year failed	Dam height m	Lake volume (m^3)	Flood peak (m^3/s)	References
Missoula	Montana, U.S.A.	16,000–12,000 year B.P.	1,078	$2,184,000 \times 10^6$	21.3×10^6	Baker 1973; Clarke et al. (1984)
Vatnsdalur	Iceland	1898	372	120×10^6	3,000	Thorarinsson (1939)
Chong Kumdan (Shyok)	India	1929	120	$1,350 \times 10^6$	22,650	Gunn (1930); Hewitt (1982)
Demmevatn	Norway	1937	406	11.6×10^6	1,000	Clague and Mathews (1973)
Graenalora	Iceland	1939	535	$1,500 \times 10^6$	5,000	Thorarinsson (1939)
Corner	Switzerland	1944	?	$>6 \times 10^6$	200	Haeberli (1983)
Gjanupsvatn	Iceland	1951	167	20×10^6	370	Amborg (1955)
Lake George	Alaska, U.S.A.	1958	40	$1,730 \times 10^6$	10,100	Stone (1963)
Tulsequah	British Columbia, Canada	1958	210	229×10^6	1,556	Marcus (1960)
Summit	British Columbia, Canada	1965	620	251×10^6	3,260	Mathews (1965
Ekalugad Valley	Baffin, Island, Canada	1967	120	4.8×10^6	200	Church (1972)
Strupvatnet	Norway	1969	186	2.6×10^6	150	Whalley (1971)
Hazard Lake	Yukon, Canada	1978	300	19.6×10^6	640	Clarke(1982)

Montana) (Bretz et al. 1956; Baker 1973); the Pleistocene overflow of Lake Bonneville at Red Rock Pass, near Preston, Idaho, around 14,000 years ago, after the breaching of a fan dam (Gilbert 1878; Malde 1968; Scott et al. 1980), etc.

Glacier dams represent a special case, with a large number of natural lakes. They are present in areas affected by continental or mountain glaciation, or even in periglacial regions, with long-lasting snows, or in areas affected by avalanches (Ballantyne and McCann 1980; Benn et al. 2006; Church 1972; Costa and Schuster 1988; Rudoy 2002a, b). Lakes can emerge behind an ice mass or alluvial accumulations. "Jökulhlaup ('glacier burst') is the Icelandic term for a flood caused by the sudden and usually catastrophic release of water impounded within or behind glacier ice" (Thorarinsson 1953). There has been important research at a global level on the formation and disappearance of lakes behind natural dams (Table 1.5): Alaska (Post and Mayo 1971); Canada (Clarke 1982; Young 1980); Iceland (Thorarinsson 1939, 1953, 1957); Norway (Aitkenhead 1960); Switzerland, Austria, France, Italy (Eisbacher and Clague 1984); Peru (Lliboutry et al. 1977);

Table 1.6 Classification of ice-dammed lakes (Blachut and Ballantyne 1976; Costa and Schuster 1988)

Type of lake	Example
A. Supraglacial	Generally small and not hazardous
B. Marginal ponded	Unnamed lake, Greenland (Sugden et al. 1985)
C. Converging ice-stream ponded	Between Lake, Axel Heiberg Island, Canada (Maag 1969)
D. Tributary stream-valley ponded	Flood Lake, British Columbia, Canada (Clarke and Waldron 1984)
E. Tributary glacier-valley ponded	Lago Rico, Argentina (Nichols and Miller 1952)
F. Interglacial ponded	Tulsequah Lake, British Columbia, Canada (Marcus 1960)
G. Dammed by tributary glacier	Shyok River Lake, Upper Indus Valley, Pakistan (Gunn 1930; Mason 1929)
H. Proglacial ice dammed	Generally small for valley glacier, can be gigantic for continental ice sheets (for example, Glaciar lake Agassiz)
I. Miscellaneous (ice-dammed craters above volcanoes; large englacial or subglacial water bodies)	Grimsvotn, Iceland (Thorarinsson 1953)

northern India (Hewitt 1982); Pakistan (Nash et al. 1985); Washington and Oregon (Richardson 1968).

The lakes formed in glacial areas vary in size, have relatively small depths and, most of the times, they have a temporary character. The "U" shaped glacial valleys may comprise a large but fine-textured natural moraine dam, with low resistance. The typology of glacier dams comprises nine categories (Table 1.6).

The most dangerous dams are part of the G category; large lakes emerge within the main valleys enclosed by tributary glaciers, with a contribution up to 100 times bigger than that of the principal glacier. Though the dams are not large, the lakes can comprise a huge volume of water (Budd 1975; Meier and Post 1969). The most interesting examples regard the mountain glaciers: on the Indus Valley in Pakistan, where 18 tributary glaciers formed a large dam (Hewitt 1982); the Chong Khumdan glacier on the Indus Valley; Shyok River, which caused a heavy flood in 1929 (Gunn 1930; Mason 1929).

"Glacier dams often fail periodically, with return periods of 1 to >10 yr. About 95 % of more than 50 jökulhlaups in the Alpes occurred from June through September, with maxima in June and August" (Tufuell 1984). "Most glacier dams are destroyed by the formation of englacial or subglacial tunnels, with up to 10 km in length" (Gilbert 1971; Nye 1976). "Another commonly evoked failure mechanism is the hydrostatic flotation hypothesis of Thorarinsson (1953), whereby subglacial drainage becomes possible when hydrostatic pressure of water in an ice-dammed lake exceeds the ice overburden pressure in an ice dam. This excess occurs when the depth of water behind an ice dam reaches 9 times the height of the ice dam" (Costa and Schuster 1988). "Other mechanisms for failure of glacier dams include (1) slow plastic yielding of ice from hydrostatic-pressure differences between the lake and adjacent, less-dense ice; (2) crack progression under combined shear stress from glacier flow and hydrostatic pressure; (3) water overflowing the ice dam and eroding a breach into the dam; (4) subglacial melting by volcanic heat; and (5) weakening of the ice dam by earthquakes" (Post and Mayo 1971). In order to avoid the catastrophe of the dambreak or melting, different methods have been applied. Unfortunately, they have proven ineffective, mostly those of dynamiting the dams (Nichols and Miller 1952).

The most important glacier dams are represented by moraine accumulations in high zones. Unfortunately, they are also the most dangerous ones, as they have caused numerous material damages or losses of human lives in the moment of destruction: Peru, British Columbia, Nepal, India, Pakistan, Russia, Canada, Austria, Argentina, etc. (Blown, Church, 1985; Burgusser et al. 1982; Clague et al. 1985; Eisbacher and Clague 1984; Galay 1985; Laenen et al. 1987; Lliboutry et al. 1977; Nolf 1966; Rabassa et al. 1979; Vuichard and Zimmermann 1987; Yesenov and Degovets 1979).

"A globally synchronous re-advance of glaciers during the last few centuries has been documented (Grove 1979) and referred to as the "Little Ice Age" (Malthes 1939) or "late-neoglacial time" (Porter and Denton 1967)". "Neoglacial time ended in the late 19th century, and since then, many mountain glaciers have retreated significantly (Porter and Denton 1967), leaving behind many moraine-dammed lakes" (Costa and Schuster 1988). The late and contemporary neoglacial time represents the cause of the most important glacier floods because the dams are young and tall, with no vegetation to stabilize them; the bank slope is big, of over 40°; the thermal processes within the ice masses determine massive melting, etc.

The large moraine dams have had a real endurance over time. Most of the times, the small moraine dams break, those made of heterogeneous, fine-textured material. If there is a breach in the moraine dams in high, isolated areas, there is not much damage because there is a low-population density in the adjacent areas (Table 1.7).

The best known incidents caused by moraine dams breaching occurred in the Andes or the Rocky Mountains: Cordillera Blanca and Cordillera Huayhuash in Peru, where the moraine dam of the Cohup Laggon caused, in 1941, a flood leading to the destruction of 50 % of the Huarez city and to the death of 6,000 people (Eisbacher 1982; Ericksen et al. 1970; Lliboutry et al. 1977); Nostetuko Lake and Klattasine Lake in British Columbia, Canada (Clague et al. 1985), etc.

The specialized literature also mentions the existence of a huge natural dam in the European north of Russia, which closed Lake Komi (Maslenikova and Mangerud 2001). The waters were contained between the ice sheet settled in the north of Europe and the mountain units within the central region. It corresponds to the Early Weichselian (80–100 ka). The lake was about 1,400 km long with a water level of about 100 m a.s.l. This lake may have had four alternative outlets: across the drainage divide towards the Volga River, leading the water southwards into the Caspian Sea; across the Polar Urals towards West Siberia; between the Barents Ice Sheet and the northern slope of the Kola Peninsula, leading the water

Table 1.7 Well-documented examples of moraine dams that have failed (Costa and Schuster 1988)

Lake or site name	Location	Date failed	Change in lake level m	Volume discharged $m^3 \times 10^4$	Flood peak m^3/s	Failure mechanism	References
Madatschfermer	Austria	Aug. 5, 1874	–	–	–	–	Eisbacher and Clague (1984)
Galritferner	Austria	Aug. 7, 1890	–	–	–	–	Eisbacher and Clague (1984)
Cohup	Peru	Dec. 13, 1941	–	–	–	–	Eisbacher, (1982); Lliboutry et al. (1977)
White Branch	Oregon, U.S.A.	July 1942	–	–	360	–	Laenen et al. (1987)
Tempanos	Argentina	1942–1953	–	–	–	Excess melt water	Rabassa et al. (1979)
Jancarurish	Peru	1950	21	600–1,000	7,000–8,000	Collapse of undercut glacier	Lliboutry et al. (1977)
Artesoncoda	Peru	July 16–17, 1951	113	–	–	Ice fall	Lliboutry et al. (1977)
Broken Top	Oregon, U.S.A.	Oct. 7, 1966	4.6	18.9	71	Ice fall	Nolf (1966); This report
Squaw Creek	Oregon, U.S.A.	Sept. 7, 1970	25.6	33.3	297	–	Laenen et al. (1987)
Safuna Alta	Peru	1970	38	490 (stored in lake)	–	Earthquake-induced piping	Lliboutry et al. (1977)
Klattasine	British Columbia, Canada	June 1971–Sept. 1973	13	170	>1,000	–	Clague et al. (1985)
Moraine no. 13	Soviet Union	Aug. 3, 1977	5.2	8.64	210	Melting of frozen soil	Yesenov and Degovets (1979)
Mingbo Valley, Dudh Kosi River	Nepal	Sept. 3, 1977	30	490	1,100	Melting of ice core? Excess runoff	Buchroithner et al. (1982); Fushimi et al. (1985)
Nostetuko	British Columbia, Canada	July 19, 1983	38.4	650	11,000	Ice fall	Blown and Church (1985)
Dig Tsho	Nepal	Aug. 4, 1985	22.5	800	1,600	Ice fall	Galay (1985); Vuichard and Zimmermann (1987)

northwestwards into the Norwegian Sea; across the drainage divide between the White Sea and the Baltic Sea catchment areas (Maslenikova and Mangerud 2001).

The moraine dams can break for various reasons: avalanches and a slope falling that create a tsunami wave that goes over the dam; excessive rains during the withdrawal of glaciers; rapid snow melting and the increase in the level of the lake water; earthquakes, etc. (Herget 2012; Herget and Agatz 2002; Plafker and Eyzaguirre 1979; Rabassa et al. 1979).

All other natural dams, with various origins, are time enduring because there are local circumstances allowing a dynamic balance of the environmental conditions. Globally, there have been a series of catastrophes because of natural dams breaching. From this perspective, there are conclusive examples on all continents. Just like in the case of artificial dams, within the ensemble natural dam/dam lake there are two types of floods: upstream or backwater flooding as the reservoir fills (on the Spanish Fork River in Utah, in April 1983) (Kaliser and Fleming 1986); downstream flooding because of failure of the dam. "Regression analysis with potential energy as the independent variable produces different equations for landslide, glacier, moraine, and earth- and rock-fill dam-failure flood peaks, with standard errors ranging from 64 % for glacier dams to 185 % for landslide dams" (Costa and Schuster 1988) (Table 1.8).

The mechanisms contributing to the destruction of natural dams are diverse and their action depends on local conditions. The most typical of them refer to piping and seepage (Table 1.9).

In case the dam is formed of a big stone block and if the lacustrine basin has a strong water balance, the lake can exist for a long period (the Red Lake, in Romania, St Andrew's Lake in France, etc.). When the lakes drain suddenly, catastrophes occur. At the end of 1840 and the beginning of 1841, there was a huge landslide that blocked the stream of the Indus River in the Nanga Parbat region of Pakistani Kashmir. In May 1841 the dam lake on the Indus was 64 km long and 300 m deep at the barrier level. When the water level touched the crest, the barrier failed and the waters were completely drained in only 24 h.

The biggest catastrophe occurred after the landslide dam on the Dadu River in Sichuan (China) failed, on June 10, 1786; then, the high- water wave propagated on 1,400 km and killed 100,000 people (Schuster and Wieczorek 2002). In 1893,

Table 1.8 Summary of regression equations to predict peak discharge from the failure of earth- and rock-fill, landslide, moraine and glacier dams (Costa and Schuster 1988)

Type of dam	Equation	Number of data points	Coefficient of determination r^2	Standard error %
1. Earth- and rock-fill	$Q = 0.0184(PE)^{0.42}$	26	0.75	91
2. Landslide	$Q = 0.0158(PE)^{0.41}$	12	0.81	185
3. Moraine	$Q = 0.00013(PE)^{0.60}$	8	0.78	92
4. Glacier	$Q = 0.0000055(PE)^{0.59}$	11	0.80	64

Note Q Peak discharge (m^3); *PE* Potential energy (joules)

Table 1.9 Predominant failure mechanisms of constructed and natural dams (Costa and Schuster 1988)

Type of dam	Predominant failure mechanism
Constructed (earth- and rock-fill)	Piping and seepage, overtopping, foundation
Glacier	Progressive enlargement of tunnels and channels
Landslide	Overtopping
Moraine	Wave overtopping

on the upper stream of the Ganga River (at the village Gohna in India), there was a huge landslide that created a 250 m deep lake. 8 months later, the water level exceeded the dam crest and the lake lowered its level by 100 m in just 2 h. The high-water wave moved forward with a 15 m height downstream, on about 100 km.

Nyos Lake is a crater lake in Cameroon; it was created after the landslide of a pyroclastic deposit, around 400 years ago. At the outlet level, the water forms a 40 m high fall. The lake is located in a volcano crater and it is 200 m deep. On August 21, 1986, carbon dioxide emerged and suffocated 1,746 villagers as they slept, at a distance of 27 km downstream. This is why over 12,000 people were evacuated from the area of the lake (Joint UNEP/OCHA Environment Unit 2005).

Betiş Lake in Romania was formed after a landslide occurred in 1956 (in the Maramureş Mountains), on the Maramureş River, a tributary of the Vaser River that was 9 m deep initially. The lake disappeared after only 2 years (Gâştescu 1979). On the Zăbala River of the Vrancea Mountains, in 1977, a natural dam lake was formed but it only lasted 3 weeks. Bolătău Lake (in the Curvature Carpathians), on the Zăbala River, tributary of the Uz River, was formed in 1883 but the initial shape and depth have altered. It is 3.5 m deep and it has changed from an initial 12 ha surface to 7 ha.

In 1978, a landslide occurred on the Cuejdul creek, in the Moldavian Subcarpathians, which created a small-sized lake. In 1991, a new, bigger landslide took place and it blocked the creek completely, thus creating the largest natural dam lake in Romania: Lacul Crucii (Cuejdul). Its surface is 12.95 ha and it registers a maximum depth of 16.3 m.

Sarez Lake in Central Asia (Tajikistan), on the Murghab River (the Pamir Plateau) was formed after a landslide caused by an earthquake (6.5–7 Richter degrees) on February 5–6, 1911. It is situated at about 3,263 m above sea level and it was 750 m deep initially. Currently, it is 505 m deep. It has a 79.7 km^2 surface, 55.8 km in length, 3.3 km maximum width, 201.8 m average depth and a 16.074 km^3 water volume. The landslide created 2.2 mln.m^3 of water and it formed a 3 km long and 550 m high dam—the highest dam in the world. Its life span is estimated at about 300 years (Bolt et al. 1975). It is the deepest natural dam lake on Earth.

The most enduring European lake is Chaillexon (Brenets), on the valley of the Doubs River (France). It dates since the Würm glacial retreat (12,000 years) and it was the effect of two landslides. The valley has been closed with extremely rough

Table 1.10 Comparison of the Mt Adams landslide dam data with a GIS model predicted landslide (Arshad et al. 2004)

Location and date of dam formation	Hancox et al. (2000)	The GIS results
Type and characteristics of the landslide		
(a) Vertical fall of debris	1,790 m	1,800 m
(b) Length of debris runout	3,000 m	2,800 m
Bedrock and surficial geology in the landslide source area	Schist and colluvium	Schist
Dam size, volume and shape		
(a) Height of landslide dam	120 m (max), 80–90 m at spillway	100 m
(b) Width of landslide dam (at spillway)	450 m	330 m
(c) Length of dam (along the valley)	650 m	700 m
(d) Volume of landslide dam	10 million m^3 ds	11 million m^3 ds
(e) Slope of dam faces (down/upstream)	24°/us 12°	24°/us 12°
Lake size, depth and volume		
(a) Length	1,200 m	1,200 m
(b) Width	350 m	250 m
(c) Maximum depth	80 m	100 m
(d) Volume	5–7 million m^3	4.5–6.5 million m^3
Estimated maximum discharge		
(a) Costa and Schuster (1988)	Upton 3,000 m^3/s	2,500 m^3/s
(b) 100 year flood	–	550 m^3/s

rocks (di-Giovanni et al. 1998). It is currently 90 % clogged. It was initially 15 km long, with 90 mln.m^3 water volume and a 750 ha surface (Bichet et al. 1995).

Clear Lake in California is the oldest, with the biggest surface (180 km^2). It was formed on the valley of the Russian River, by damming with volcanic lava, with a subsequent landslide, which separated Clear Lake from Blue Lake. It is 31 km long, 13 km wide (maximum), 8.2 m deep (average), with a maximum depth of 18 m. It is around 480,000 years old.

Natural dams can last from a couple of minutes to thousands or tens of thousands of years. The longevity of dams and of natural dam lakes, implicitly, depends on several factors. These are: the volume of material comprising the dam, the water volume accumulated behind the dam, the size and shape of the dams and lakes, the triage of the materials, the water loss ratio, the sedimentation ratio, etc.

Because the new society requires an increased protection of special natural areas and, at the same time, the elimination of natural hazards, proper management is mandatory for each unit. Firstly, the shapes and morpho-bathymetric parameters of the dams and of the lake basins should be inventoried. The data obtained through Geographic Information System (GIS) clearly demonstrate that all the old parameters, obtained through classical methods, must be re-assessed (Table 1.10).

Such lakes may also have practical purposes, besides the tourist ones. Waikaremoana Lake in New Zealand, situated behind a 250 m high dam, was the basis for a hydroelectric power station. In this regard, between 1935 and 1950, a tunnel evacuating the excess lake water, thus setting up a micro-power station (Offer 1997), penetrated the natural dam.

References

Adams J (1981) Earthquake-dammed lakes in New Zealand. Geology 9:215–219

Aitkenhead N (1960) Observations on the drainage of a glacier-dammed lake in Norway. J Glaciol 3:607–609

Alden WC (1928) Landslide and flood at Gros Ventre, Wyoming. Am Inst Min Metall Eng Trans 76:347–358

Amborg L (1955) Hydrology of the glacial river Austurfljot. Geogr Ann 36(3–4):185–201

Anderson GP (1948) Waikaremoana—the problem of lake control. New Zealand Inst Eng Proc 34:508–526

Antonescu CS (1967) Biologia apelor. Didactical and Pedagogical Publishing House, Bucharest

Arlinghaus R, Johnson BM, Wolter C (2008) The past, present and future role of limnology in freshwater fisheries science. Intern Rev Hydrobiol 93(4–5):541–549

Arshad WA, Doscher C, Davies TRH (2004) Landslide dambreak hazard analysis in Westland. In: Liong SY, Phoon KK, Babovic V (eds) The 6th international conference on hydroinformatics, 2nd edn. World Scientific, Singapore, pp 730–737, doi: http://books.google.ro/books?id=-c8tizA5fdoCpg=PA730lpg=PA730dq=Landslide+dambreak+hazard+analysis+in+Westlandsource=blots=Qob0eCKzCEsig=JOBXoGWsqomT27mwbNUCR6l1Pmwhl=ro&sa=Xei=cFJ1Ue7KJ-Sk4gT4kYDQBQved=0CDQQ6AEwAQ#v=onepageq=Landslide%20dambreak%20hazard%20analysis%20in%20Westlandf=false

Baker VR (1973) Paleohydrology and sedimentology of Lake Missoula flooding in eastern Washington. Geol Soc Am Spec Pap 144:1–79

Baker VR, Costa JE (1987) Flood power. In: Mayer L Nash D. (eds) Catastrophic flooding, 1–21, Allen and Unwin, Boston

Ballantyne CK, McCann SB (1980) Short-lived damming of a high Artic ice-marginal stream, Ellesmere Island, N.W.T., Canada. J Glaciol 25:487–491

Benn DI, Owen LA, Finkel RC, Clemmens S (2006) Pleistocene lake outburst floods and fan formation along the eastern Sierra Nevada, California: implications for the interpretation of intermontane lacustrine records. Quatern Sci Rev 25:2729–2748

Berthling I, Sollid JL (1999) The drainage history of glacial lake Nedre Glåmsjø, southern central Norway. Norwegian J Geog 53(4):190–201, Rutledge Publisher

Bichet V, di-Giovanni C, Richard H, Campy M (1995) Changing patterns of material flow since 12000 BP in the upper Doubs valley (France). Quaternaire 6(2):85–90

Blachut SP, Ballantyne CK (1976) Ice-dammed lakes-A critical review of their nature and behavior. Hamilton, Ontario, McMaster University department of Geography, Discussion 6:1–90

Blown I, Church M (1985) Catastrophic lake drainage within the Homathko River basin British Columbia. Can Geotech J 22(4):551–563

Bolt BA, Horn WL, Macdonald GA, Scott RF (1975) Geological hazards: earthquakes, tsunamis, volcanoes, avalanches, landslides, floods. Springer, New York

Breier A (1976) Lacurile de pe litoralul românesc al Mării Negre: Hydrogeographic study. Romanian Academy Publishing House, Bucharest

Bretz JH, Smith HTU, Nelf GE (1956) Channeled scablands of Washington: new data and interpretation. Geol Soc Am Bull 67:957–1049

Buchroithner MF, Jentsch G, Wanivenhaus B (1982) Monitoring of recent geological events in the Khumbu area (Himalaya, Nepal) by digital processing of Landsat MSS data. Rock Mech 15:181–197

Budd WF (1975) A first simple model for periodically self-surging glaciers. J Glaciol 14:3–21

Burgusser HM, Gansser A, Pika J (1982) Late glacial lake sediments of the Indus Valley area, northwestern Himalaya. Ecologae Geol Helvetae 75:51–63

Butler DR, Oelfke JG, Oelfke LA (1986) Historic rockfall avalanches, northeastern Glacier National Park, Montana, U.S.A. Mt Res Dev 6:1–11

Chang L (1934) Investigation of the deixi Earthquake, Sichuan. The Geological Society of China, Nanking, China. Comments on Geology 3(3):251–292

Chang SC (1984) Tsao-Ling landslide and its effect on a reservoir project. In: Proceedings of the 4th international symposium on Landslides, Toronto, Sep 16–21, vol 1, pp 469-473

Church M (1972) Baffin Island sandurs: a study of Arctic fluvial processes. Geol Soc Can Bull 216:208

Clague JJ, Mathews WH (1973) The magnitude of jökulhlaups. J Hydrol 12:501–504

Clague JJ (1973) Late Cenozoic Geology of the Southern Rocky Mountain Trench, British Columbia. Ph. D. Thesis. Department of Geological Sciences, University of British Columbia

Clague JJ, Evans SG, Blown IG (1985) A debris flow triggered by the breaching of a moraine-dammed lake, Klattasine Creek, British Columbia. Can J Earth Sci 22:1492–1502

Clague JJ, Evans SG (1994) Formation and failure of natural dams in the Canadian Cordillera. Geological Survey of Canada, Bulletin 464, Natural Resources Canada, Ottawa, Ont

Clark TH (1947) Two recent landslides in Quebec. Trans R Soc Can 41(3–4):9–18

Clarke GKC (1982) Glacier outburst floods from "Hazard Lake", Yukon territory, and the problem of flood magnitude prediction. J Glaciol 28:3–21

Clarke GKC, Waldron DA (1984) Simulation of the August 1979 sudden discharge of glacier-dammed Flood Lake, British Columbia. Can J Earth Sci 21:502–504

Cohen AS (2003) Paleolimnology: The history and evolution of lake systems. Oxford University Press, Oxford

Collet LW (1925) Las lacs, leur mode de formation, leurs eaux, leurs destin. Editions Gaston Doin, Paris

Conea I (1965) Margini şi Mărgineni. Geogr Commun Bucharest 3:275–286

Costa JE, Schuster RL (1988) The formation and failure of natural dams. Geol Soc Am Bull 100:1054–1068

Cotecchia V (1978) Systematic reconnaissance mapping and registration of slope movements. Bull Int Assoc Eng Geol 17:5–37

Cotton CA (1941) Landscape, as developed by the processes of normal erosion. Wiley, New York 509

Coxon P, Owen LA, Mitchell WA (1996) A late Quaternary catastrophic flood in the Lahul Himalayas. J Quat Sci 11(6):495–510

Crandell DR, Varnes DJ (1961) Movement of the Slumgullion earth-flow near |Lake City, Colorado. U.S. Geological survey professional paper, 424-B, pp 136–139

Cruden DM (1982) The Brazeau Lake slide, Jasper National Park, Alberta. Canadian J Earth Sci 19(5):975–981

Davis WM (1882) On the classification of lake basins. Boston Society of Natural History Proceedings 21:315–381

Davis WM (1933) Granite domes of the Mojave desert, California. Trans San Diego Soc Nat Hist 7:211–258

Davis WM (1887) On the classification of lake basins. Science 10:142–155

Dawson GM (1898) Remarkable landslip in Portneuf County, Quebec. Geol Soc Am Bull 10:484–490

Delebecque A (1898) Les Lacs Francais. Chamerot et Renouard, Paris

di-Giovanni C, Disnar JR, Bichet V, Campy M, Guillet B (1998) Geochemical characterization of soil organic matter and variability of a postglacial detrial organic supply (Chaillexon lake, France). Earth Surf Process Land 23(12):1057–1069

Dong JJ, Tung YH, Chen CC, Liao JJ, Pan YW (2011) Logistic regression model for predicting the failure probability of a landslide dam. Eng Geol 117:52–61
Dussart B (1966) Limnologie: L'etude des eaux continentales. Gauthier-Villars, Paris 678
Dussart B (1992) Limnologie, l'étude des eaux continentales. Boubée, Paris
Eisbacher GH (1982) Mountain torrents and debris flows. Episodes 4:12–17
Eisbacher GH, Clague JJ (1984) Destructive mass movements in high mountains—Hazard and management. Geol Surv Can Pap 84(16):230
Emerson FB (1925) 180-ft. dam formed by landslide in Gros Ventre Canyon. Eng News-Rec 95(12):467–468
Ericksen GE, Plafker G, Concha JF (1970) Preliminary report on the geologic events associated with the May 31, 1970, Peru earthquake, vol 639. U.S. Geological Survey Circular, Virginia, p 25
Evans SG (1984) The 1880 landslide dam on Thompson River near Ashcrof, British Columbia. Curr Res: Geol Surv Can Pap 84(1A(A)):655–658
Fahnestock RK, Bradley WC (1973) Knik and Matanuska Rivers, Alaska: a contrast in braiding. In: Morisawa ME (ed.) Fluvial geomorphology, State University of New York, Binghamton, 220–250
Finch RH (1937) A tree-ring calendar for dating volcanic events: Cinder Cone, Lassen National Park, California. Am J Sci 33:140–146
Forbes SA (1887) The Lake as a Microcosmos. Bull Peoria Ill Sci Assoc, Reprint 1925 in Bull Ill Nat Hist Surv, p 15
Forel FA (1892) Le Lehman: monographie limnologique. Tome I, F.Rouge, Lausanne
Forel FA (1895) Le Lehman: monographie limnologique. Tome II, F.Rouge, Lausanne
Forel FA (1904) Le Lehman: monographie limnologique. Tome III, F.Rouge, Lausanne
Fushimi H, Ikegami K, Higuchi K, Shankar K (1985) Nepal case study: catastrophic floods. In: Young GJ (ed) Techniques for prediction of runoff from glacierized areas, vol 149. International Association of Hydrological Sciences, Wallingford, pp 125–130
Galay VJ (1985) Hindu Kush-Himalayan erosion and sedimentation in relation to dams. Chinese Academy of Science. In: Proceeding of the international workshop on water management in the Hindu Kush-Himalaya region, Chendu, China, p. 24
Gallino GL, Pierson TC (1985) Polallie Creek debris flow and subsequent dam-break flood of 1980, East Fork Hood River basin, Oregon, vol 2273. U.S. Geological Survey, Water Supply Paper, Virginia, p 22
Gasiev E (1984) Study of the Usoy landslide in Pamir. In: Proceedings of the 4th international symposium on landslides, Toronto, Canada, 16–21 Sep 1984, vol 1, pp 511–515
Gâştescu P (1971) Lacurile din Romania—Limnologie regionala. Romanian Academy Publishing House, Bucharest (in Romanian), pp 1–372
Gâştescu P (1979) Lacurile Terrei. Albatros Publishing, Bucharest
Gâştescu P (1998) Hidrologie. Roza vânturilor Publishing House, Târgovişte
Geikie J (1905) Structural and field geology for students of pure and applied science. Oliver and Boyd, London
Gilbert GK (1878) The ancient outlet of Great Salt Lake. Am J Sci 3(15):256–259, doi: http://maps.nls.uk/dcn6/7443/74432608.6.pdf
Gilbert R (1971) Observations on ice-dammed Summit Lake, British Columbia, Canada. J Glaciol 10:351–356
Glade T, Crozier MJ (1999) Landslides in New Zealand: a selected bibliography—research report. School of Earth Science, Wellington, pp 1–103
Glazzyrin GY, Reyzvikh VN (1968) Compulation of the flow hydrograph for the breach of landslide lakes. Sov Hydrol 5:592–596
GRAMMONT (1983) Dictionaire Encyclopédique Alpha. Editions GRAMMONT, Lausanne
Grove JM (1979) The glacial history of the Holocene. Prog Phys Geogr 3:1–54
Guilcher A (1979) Précis d'hydrologie (marine et continentale). Editions Masson, Paris
Gunn JP (1930) The Syook flood, 1929. Himalayan J 2:35–47

Hadley JB (1964) Landslides and related phenomena accompanying the Hebgen Lake earthquake of August 17, 1959. U.S. Geol Sur Prof Paper 435:107–138
Haeberli W (1983) Frequency and characteristics of glacier floods in the Swiss Alps. Ann Glaciol 4:85–90
Halbfass W (1903)—se sterge
Halbfass W (1923) Grundzüge einer vergleichenden Seenkunde. Borntraeger, Berlin, pp 1–354
Hancox GT, McSaveney MJ, Davies TR, Hodgson DR (2000) The October 1999 landslide dam in Poerua River, Westland, New Zealand. NZSOLD Symposium, November 2000
Herdendorf CC (1990) Great Lakes Estuaries. Estuaries 13(4):493–503
Herget J (2012) Ice-dammed lake outburst floods in the Altai Mountains, Siberia: a review with links for further readings. Tomsk State Univ J Biol 1(17):148–168
Herget J, Agatz H (2002) Modelling ice-dammed lake outburst floods in the Altai Mountains (Siberia) with HEC-RAS. In: Thorndycraft VR, Benito G, Barriendos M, Llasat MC (eds) Palaeofloods, historical floods and climatic variability: application in flood risk assessment. Proceedings of the PHEFRA Workshop, Barcelona, 16–19 Oct 2002
Herget J, Carling PA (2004) Review on large scale gravel dunes causes by Pleistocene ice-dammed lake outburst floods. Marine Sandwave and River Dune Dynamics, Enschede, the Netherlands, pp 96–101
Hewitt K (1982) Natural dams and outburst floods of the Karakoram Himalaya. In: Glen JW (ed) Hydrological aspects of alpine and high-mountain areas, vol 138. International Association of Hydrological Sciences Publication, UK, pp 259–269
Holmes A (1965) Principles of physical geology (2nd edition). Ronald Press, New York, pp 1–1288
Huss M, Bauder A, Werder M, Funk M, Hock R (2007) Glacier-dammed lake outburst events of Gornersee Switzerland. J Glaciol 53(181):189–200
Hutchinson GE (1957) A treatise on limnology, I. Geography, physics and chemistry. Wiley, New York, pp 1–1015
Hutchinson JN, Kojan E (1975) The Mayunmarca landslide of 25 April 1974, vol Serial 3124/RHO.RD/SCE. UNESCO, Paris, pp 1–23
Imboden C (1976) Eaux vivantes: Initiation a la biologie des zones humides. Editions Ligne Suisse pour la Protection de la Nature, Bale
Jakobson M, Björck S, Alm G, Andren T, Lindeberg G, Svensson NO (2007) Reconstructing the Younger Dryas ice dammed lake in the Baltic Basin: Bathymetry, area and volume. Glob Planet Changes 57:355–370
Jansen RB (1988) Advanced dam engineering for design, construction, and rehabilitation. Kluwer Academic Publishers, Netherlands
Joint UNEP/OCHA Environment Unit (2005). Lake Nyos Dam Assessment. http://ochaonline.un.org/ochaunep
Jones DP, Ricker KE, DeSloges JR, Maxwell M (1985) Glacier outburst flood on the Noeick River—the draining of Ape Lake, British Columbia, October 20, 1984. Geol Surv Can Open-File Rep 1139:1–81
Kaliser BN, Fleming RW (1986) The 1983 landslide dam at Thistle, Utah. In: Schuster RL (ed) Landslide dams: processes, risk, and mitigation, vol 3. American Society of Civil Engineers Geotechnical Special Publications, Virginia, pp 59–83
Kargel JS, Leonard G, Crippen RE, Delaney KB, Evans SG, Schneider J (2010) Satellite Monitoring of Pakistan's Rockslide-Dammed Lake Gojal. Eos, 91(43):394–395
King WDVO (1934) The Mendoza River flood of 10–11 January 1934, Argentina. Geogr J 84:321–326
Knight DK, Bennett PT (1960) Stability of slide dam and recommendations on development of overflow spillway. In: Madison River, Montana, report on flood emergency, Madison River slide. U.S. Army Corps of Engineers, Riverdale, North Dakota, Appendix, **VI(5)**
Laenen A, Scott KM, Costa JE, Orzol LL (1987) Hydrologic hazards along Squaw Creek from a hypothetic failure of the glacial moraine impounding Carver Lake near Sisters, Oregon. U.S. Geological Survey Open-File, Report 87–41:1–48

Lee KL, Duncan JM (1975) Landslide of April 25, 1974, on the Mantaro River Peru. Committee on Natural Disasters, Commission on Sociotechnical System, National research Council, Washington DC, pp 1–72

Li T, Schuster RL, Wu J (1986) Landslide dams in south-central China. In: Schuster RL (ed) Landslide dams—processes, risk, and mitigation, vol 3. American Society of Civil Engineers Geotechnical, Special Publication, Virginia, pp 146–162

Lliboutry L, Morales B, Poutre A, Schneider B (1977) Glaciological problems set by the control of dangerous lakes in Cordillera Blanca, Peru. I. Historical failures of morainic dams, their causes and preventions. J Glaciol 18(79):239–254

Loope DB, Swinehart JB, Mason JP (1995) Dune-dammed paleovalleys of the Nebraska Sand Hills: Intrinsic versus climatic controls on the accumulation of lake and marsh sediments. GSA Bull 107(4):396–406

Maag HV (1969) Ice-dammed lakes, and marginal glacial drainage on Axel Heiberg Island. Montreal, Axel Heiberg Island Research Report, McGill University, pp 1–147

Malde E (1968) The catastrophic late Pleistocene Bonneville flood in the Snake River plain, Idaho. U.S. Geol Surv Prof Pap 596:1–52

Malthes FE (1939) Report of Committee on Glaciers. Am Geophys Union Trans 20:518–523

Marcus MG (1960) Periodic drainage of glacier-dammed Tulsequah Lake, British Columbia. Geogr Rev 50:89–106

Marr JE (1916) The geology of the Lake District and the scenery as influenced by geological structure. Cambridge University, Cambridge, pp 1–376

Marsh P, Neumann N (2003) Climate and hydrology of a permafrost dammed lake in NW Canada, Permafrost. In: Springman SM, Arenson LU (eds) Phillips M. Swets and Zeitllinger, Lisse, pp 729–734

de Emm M (1902) La Valachie. Essai de monographie geographique, Paris

de Emm M (1909) Traité de Géographie Physique. A. Colin, Paris

Maslenikova O, Mangerud J (2001) Where was the outlet of the ice-dammed Lake Komi, Northern Russia? Glob Planet Change 31:337–345

Mason K (1929) Indus floods and Shyok glaciers. Himalayan J 1:10–29

Mathews WH (1965) Two self-dumping ice-dammed lakes in British Columbia. Geogr Rev 55:46–52

Mayo LR (1987) Cause of the advance of Hubbard Glacier, Alaska [abs.]. In: Alaska's resources, Alaska's future: Proceedings, Alaska Science Conference, September 1987, 49

Mayo LR (1989) Advance of Hubbard Glacier and 1986 outburst of Russell Fiord, Alaska, U.S.A. Annals of Glaciology 13:189–194

Mehedinţi S (1930) Terra: Introducere în geografie ca ştiinţă. "S.Ciornei" National foundation Publishing House, Bucharest

Meier MA, Post A (1969) What are glacier surges? Can J Earth Sci 6:807–817

Meybeck M (1995) Global distribution of lakes. In: Lerman A, Imboden DM, Gat JR (eds) Physics and chemistry of lakes, Springer-Verlag: 1–35

Meyer GA, Leidecker ME (1999) Fluvial terraces along the Middle Fork Salmon River, Idaho, and their relation to glaciation, landslide dams, and incision rates: a preliminary analysis and river-mile guide. In: Huges SS, Thackray GD (eds) Guidebook to the geology of Eastern Idaho: Pocatello. Idaho Museum of Natural History, Idaho, pp 219–235

Meyer W, Sabol MA, Schuster RL (1986) Landslide dammed lakes at Mount St. Helens, Washington. In: Schuster RL (ed) Landslide dams-Processes, risk, and mitigation, vol 3. American Society of Civil Engineers Geotechnical Special Publications, Virginia, pp 21–41

Mindrescu M, Cristea IA, Florescu G (2010) Water quality and ecology of the Iezer and Bolătău Lakes. Lakes, reservoirs and ponds 4(2):119–130

Murray J (1900) Barymetric Survey of Scottish Fresh Water Lochs of Scotland. Challenger Office, Edinburgh

Nash DFT, Brunsden DK, Hughes RE, Jones DKC, Whalley BF (1985) A catastrophic debris flow near Gupis, northern areas, Pakistan. In: Proceedings of the 11th international conference

on soil mechanics and foundation engineering, San Francisco, 12–16 Aug 1985, vol 3, pp 1163–1166
Neuffer DP, Bruhn RL (2005) Landslide investigation of timber lakes estates, Wasatch County, Utah: landslide inventory and preliminary geotechnical-engineering slope stability analysis. Utah Geological Survey, Salt Lake City, pp 1–65
Nichols RL, Miller MM (1952) The Moreno Glacier, Lago Argentino, Patagonia, advancing glaciers and nearby simultaneous retreating glaciers. J Glaciol 2:41–50
Nolf B (1966) Broken top breaks—flood released by erosion of glacial moraine. The Ore Bin 28(10):182–188
Nye JF (1976) Water flow in glaciers: jokulhlaups, tunnels and veins. J Glaciology 17(76):181–207
Noord H, Phuntsho T (2005) Tsatichu Revisited: Hazard appraisal after the breach of the landslide dammed lake and some comments on mitigation options. Department of Geology and Mines
O'Connor JE, Baker VR (1992) Magnitudes and implications of peak discharges from glacial Lake Missoula. Geol Soc Am Bull 104:267–279
Offer RE (1997) Walls for water: pioneer dam building in New Zealand. The Dunmore Press Ltd, Palmerston North
Ousley R (1788) An account of the moving of a bog, and the formation of a lake, in the county of Galway, Ireland. Trans R Ir Acad 2:3–5
Penck A (1882) Die Vergletscherung der deutschen Alpen. Johann Ambrosium Barth, Leipzig (Barth)
Pisota I, Nastase A (1957) Lacul Rosu, nod de confluenta a trei bazine hidrografice. Probl De Geogr (in Romanian) 4:181–205
Pişota I (1971) Lacurile glaciare din Carpaţii Meridionali. Romanian Academy Publishing House, Bucharest
Pişota I (1995) Hidrologie. University Publishing House, Bucharest
Pişota I, Buta I (1983) Hidrologie. Didactical and Pedagogical Publishing House, Bucharest
Plafker G, Eyzaguirre VR (1979) Rock avalanche and wave at Chungar, Peru. In: Voight B (ed) Rockslides and avalanches, vol 2. Elsevier, Amsterdam, the Netherlands, pp 269–279
Porter SC, Denton GH (1967) Chronology of neoglaciation in the North American Cordillera. Am J Sci 267:177–210
Post A, Mayo LR (1971) Glacier dammed lakes and outburst floods in Alaska. U.S. Geological Survey Hydrologic Investigations Atlas HA-455, Virginia, pp 1–10
Pusharenko VP (1982) Regularities of mud-rock flow formation in central Asia. In: Sheko A (ed) Landslides and mudflows reports of Alma-Ata International Seminar, U.S.S.R. Centre of International Projects, GKNT, Moscow, pp 87–106
Rabassa J, Rubulis S, Suarez J (1979) Rate of formation and sedimentology of (1976–1978) push-moraines, Frias Glacier, Mount Tronador (41°10′ S; 71°53′ W), Argentina. In: Schluster Ch (ed) Moraines and varves. Rotterdam, the Netherlands, A.A. Balkema, pp 65–79
Richardson D (1968) Glacier outburst floods in the Pacific Northwest. US Geol Surv Prof Pap 600:79–86
von Richtofen F (1886) Fuhrer fur Forschungsreisende. Janecke, Hannover, pp 1–734
Romanescu G (2003) Hidrologie generală. Terra Nostra Publishing House, Iaşi, pp 1–574
Romanescu G (2005) Ţara de Foc şi ultimul ei conchistador. Didactical and Pedagogical Publishing House, Bucharest
Romanescu G, Jigau G (1998) Geomorfologie. Universitatea de Stat din Moldova, Chisinau
Rudoy AN (2002a) Glacier-dammed lakes and geological work of glacial superfloods in the Late Pleistocene, Southern Siberia, Altai Mountains. Quatern Int 87:119–140
Rudoy AN (2002b) Late-Quaternary diluvial floodstreams in the mountains of Altai and Tuva (paleoglaciological base, geological effect, level of problem study). Lunar Planet Sci, vol 33
Russell IC (1893) A geological reconnaissance in central Washington. U.S. Geological Survey Bulletin 108:168
Russell IC (1895) Present and extinct lakes of Nevada. Nat Geogr Monogr 1(4):101–136

Săndulache A (1970) Lacurile dulci din Câmpia Transilvaniei. Ph. D thesis, Pedagogical Institute, Oradea
Seitz HR, Thomas DS, Tomlinson B (1986) The storage and release of water from a large glacier-dammed lake: Russell Lake near Yakutat, Alaska, U.S. Geological Survey Open-File Report 86–545, p 10
Schneider JF, Gruber FE, Mergili M (2011) Recent cases and geomorphic evidence of landslide-dammed lakes and related hazards in the mountains of Central Asia. Proc Second World Landslide Forum, Rome, pp 1–7
Schneider JL, Pollet N, Chapron E, Wessels M, Wassmer P (2004) Signature of Rhine Valley sturzstrom dam failures in Holocene sediments of Lake Constance, Germany. Sed Geol 169:75–91
Schram M (1968) Câteva aspect privind bilanţul hidrologic al lacurilor din Câmpia Moldovei: Geografie-Geologie. Annals of the "Al.I. Cuza" University, Iaşi, p XIV
Schuster RL, Costa JE (1986) A perspective on landslide dams. American Society of Civil Engineers, Virginia
Schuster RL, Evans SG (2011) Engineering measures for the Hazard reduction of landslide dams. In: Evans SG, Hermanns RL, Strom A, Scarascia-Mugnozza G (eds) Natural and Artificial Rockslide Dams. Springer, Berlin, pp 1–642 133, 77-100
Schuster RL, Wieczorek GF (2002) Landslide triggers and types, In: Landslides: Proceedings of the First European Conference on Landslides. Balkema Publishers, Prague
Scott RG (1970) Landslides in Cache Canyon downstream from the Wilson Valley dam sites: State of California. The Resources Agency, Department of Water Resources, Northern District
Scott WE, Meloy WD, Shroba RR, Miller RD (1980) New interpretation of the late Quaternary history of Lake Bonneville, western United States. Am Quat Assoc Abstr Programs 6:168
Silva L, Cocheme JJ, Duffield WA, Tailing RI (1982) The March-April 1982 eruption on Chichoaal Volcano, Chiapas, Mexico—Preliminary observations (abs.). EOS (American Geophysical Union Transactions), vol 63, p 1126
Snow DT (1964) Landslide of Cerro Condor-Sencca, Department of Ayacucho, Peru. In: Kiersch GA (ed) Engineering geology case histories. Geological Society of America, pp 1–6
Stone KH (1963) The annual emptying of Lake George, Alaska. Arctic 16:26–40
Sugden DE, Clapperton CM, Knight PG (1985) A jökulhlaup near Sondre Stromfjord, West Greenland, and some effects on the ice-sheet margin. J Glaciol 31:366–368
Swanson FJ, Oyagi N, Tominaga M (1986) Landslide dams in Japan. In: Schuster RL (ed) Landslide dams: processes, risk, and mitigation, vol 3. American Society of Civil Engineers Geotechnical Special Publications, Virginia, pp 131–145
Thienemann AF (1925) Die Binnengewässer Mitteleuspas: Eine Limnologische Einführung. Verlag Schweizerbart, Germany
Thienemann AF (1926) Der Nahrungskreislauf Im Wasser. Verh Dtsch Zool Ges 31:29–79
Thorarinsson S (1939) The ice-dammed lakes of Iceland with particular reference to their value as indicators of glacial oscillations. Geogr Ann 21(3–4):216–242
Thorarinsson S (1953) Some new aspects of the Grimsvotn problem. J Glaciol 2:267–275
Thorarinsson S (1957) The jökulhlaup from the Katla area in 1955 compared with other jökulhlaups in Iceland, Reykjavik. Museum Nat Hist Misc Pap 18:21–25
Tómasson H (2002) Catastrophic floods in Iceland. In: Snorrason A, Finnsdóttir HP, Moss ME (eds) The extremes of the extremes: extraordinary floods, 121–128, IAHS Press, Centre for Ecology and Hydrology, Wallingford, Oxfordshire, UK
Tufuell L (1984) Glacier hazards. Longman, London, pp 1–97
Vuichard D, Zimmermann M (1987) The 1985 catastrophic drainage of a moraine-dammed lake, Khumbu Himal: cause and consequences. Mt Res Dev 7:91–110
Walder JS, Costa JE (1996) Outburst floods from glacier-dammed lakes: the Effect of mode of lake drainage on Flood magnitude. In: Earth Surface Processes and Landforms 21:701–723
Wetzel R (2001) Limnology: lake and river ecosystems, 3rd edn. Elsevier

Whalley WB (1971) Observations of the drainage of an ice-dammed lake—Strupvatnet, Troms, Norway. Nor Geogr Tidsskr 25(3–4):165–174
WMO-UNESCO (1992) International glossary of hydrology. WMO-UNESCO, Geneve-Paris
Yesenov UY, Degovets AS (1979) Catastrophic mudflow on the Bol'shaya Almatinka River in 1977. Soviet Hydrol 18:158–160
Young GJ (1980) Monitoring glacier outburst floods. Nord Hydrol 11:285–300
Zăvoianu I (1999) Hidrologie. România de Mâine Foundation Publishing House, Bucharest

Chapter 2
Geographic Location and Boundaries

Abstract The Red Lake is situated in the Central Group of the Eastern Carpathians, within the Hăghimaş Mountains (Hăşmaşu Mare). The Hăghimaş Mountains are part of the Moldavian-Transylvanian Carpathians group. They correspond to the southern sector of the External Marginal Syncline, with a median position between a volcanic chain in the west and a flysch chain in the east (Cristea in Munţii Hăşmaş şi Staţiunea Lacu Roşu. Sport-Turism Publishing House, Bucharest, 1978; Mihăilescu in Carpaţii Sud-estici. Scientific Publishing House, Bucharest, 1963; Niculescu and Oancea in Munţii Giurgeului şi Curmăturii. Geografia României, vol.III. Romanian Academy Publishing House, Bucharest, 1987). The name "red" comes from the fact that, at sunrise, the sun rays fall directly on the reddish clays of the western slope (Piciorul Licoş), which are reflected in its relatively clear waters. The most important tributaries are the Oaia brook (also known as Oii) and the Sec brook on the right side; on the left, there are the Vereşcheu, Licoş and Suhard brooks.

The Red Lake is situated in the Central Group of the Eastern Carpathians, within the Hăghimaş Mountains (Hăşmaşu Mare). The Hăghimaş Mountains are part of the Moldavian-Transylvanian Carpathians group. They correspond to the southern sector of the External Marginal Syncline, with a median position between a volcanic chain in the west and a flysch chain in the east (Cristea 1978; Mihăilescu 1963; Niculescu and Oancea 1987). The Red Lake lays on the middle stream of the Bicaz River, upstream from its point of confluence with Bicăjel (Fig. 2.1).

As regards the latitude, it is located between 46°47′0″ N latitude in the southern sector and 46°47′37″ N latitude in the northern sector. Longitudinally, it is situated between 25°47′0″ E longitude in the northwestern sector and 25°47′30″ E longitude in the eastern sector.

G. Romanescu et al., *Limnology of the Red Lake, Romania*,
DOI: 10.1007/978-94-007-6757-7_2,

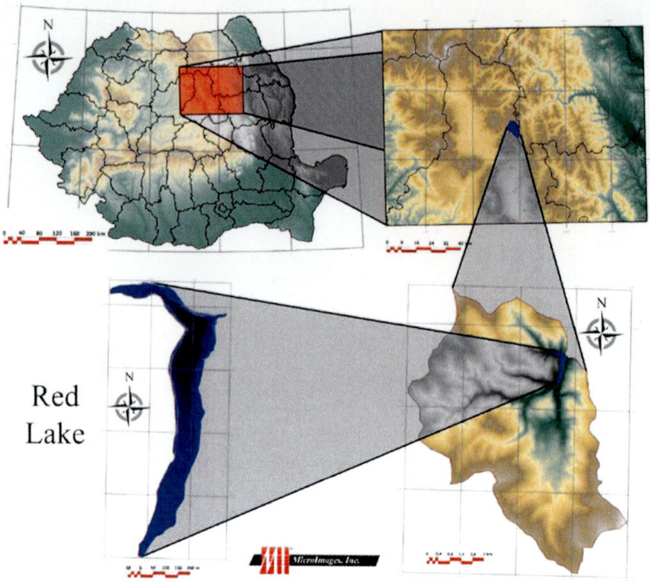

Fig. 2.1 Geographic location of the Red Lake on the territory of Romania

The name "red" comes from the fact that, at sunrise, the sun rays fall directly on the reddish clays of the western slope (Piciorul Licoş), which are reflected in its relatively clear waters.

Its main tributaries are the Oaia brook (also known as Oii) and the Sec brook on the right side; on the left, there are the Vereşcheu, Licoş and Suhard brooks. The most important ones, with regards to discharge, are the Oaia (Oii) and Suhard brooks (Fig. 2.2).

The Hăghimaş Mountains have Valea Bistricioarei as a northern limit and Mihăileni-Valea Frumoasei in the south. The conterminous landforms are the Bistriţei Mountains in the north, the Giurgeului Mountains in the south, and the Ciucului Depression in the west as well as the Ceahlău and Tarcăului Mountains in the east and south. They are 50 km long with a north–south alignment. The Red Lake hydrographic basin is surrounded by the following mountains: Piciorul Licaş (in the west), Suhardul Mic (in the north), Bâtca lui Cioflec (in the south) and Ucigaşu (in the east) (Fig. 2.3).

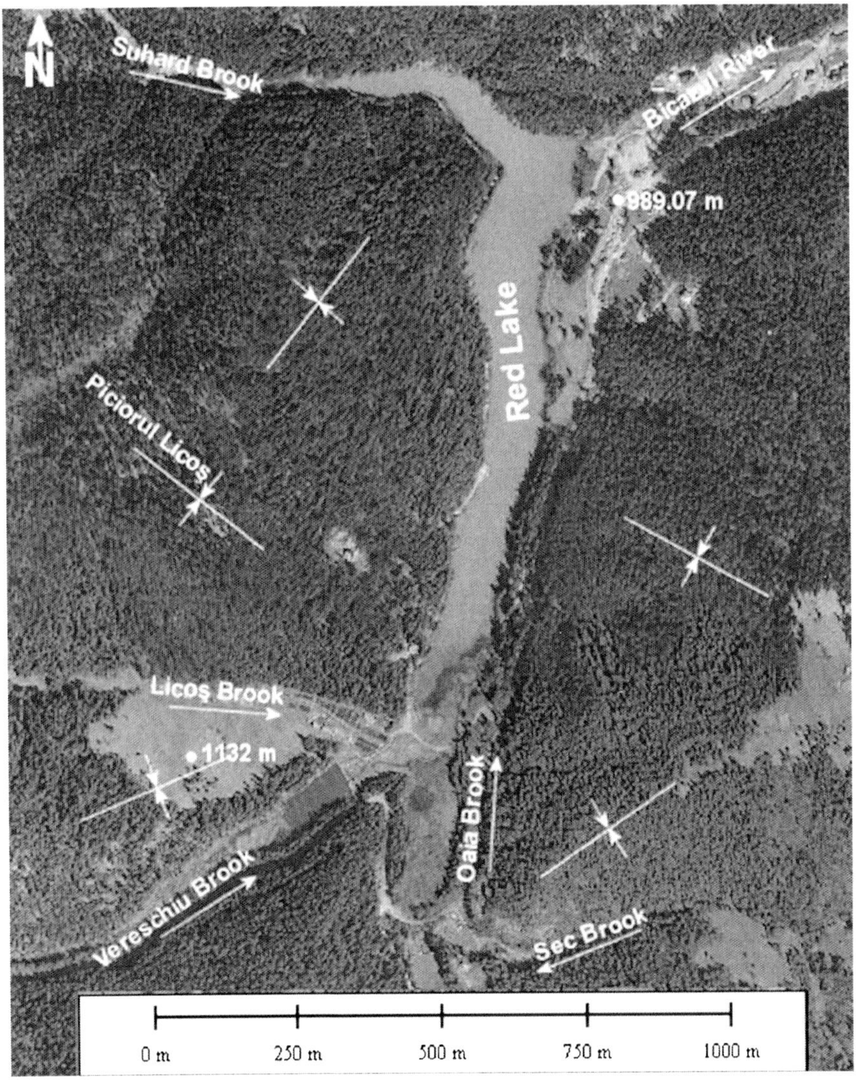

Fig. 2.2 The geographic location of the Red Lake on an aerial image (ortophotoplan www.ancpi.ro)

Fig. 2.3 The main landforms and the most imposing peaks of the Hăghimaş Mountains

References

Cristea E (1978) Munţii Hăşmaş şi Staţiunea Lacu Roşu. Sport-Turism Publishing House, Bucharest

Mihăilescu V (1963) Carpaţii Sud-estici. Scientific Publishing House, Bucharest

Niculescu G, Oancea D (1987) Munţii Giurgeului şi Curmăturii. Geografia României, vol.III. Romanian Academy Publishing House, Bucharest

Chapter 3
The Red Lake Within the Bicaz Gorges-Hăghimaş National Park

Abstract The Red Lake is part of the Bicaz Gorges-Hăghimaş National Park. It is situated in the Group of Giurgeului and Hăghimaş Mountains (Curmăturii or Hăşmaş) within the Moldavian-Transylvanian Carpathians (Romanian Carpathians and Transylvanian Depression 1987). It must be mentioned that the delimitation of mountain units has been an intensely disputed subject within the past few years, the reason for which the limits of the Hăghimaş Mountains or even its name are still being questioned. The Bicaz Gorges-Hăghimaş National Park has objectives related to hydrology, hydrogeology, flora, fauna, archaeology and to the geologic substrate. Besides the lake basin being occupied by the Red Lake and by karstic forms dominated by the Bicaz Gorges, there are also numerous geologic sites such as fossil-bearing points with a biostratigraphic and paleontologic value, a series of petrographic outcrops with paleogeographic significance and some structural perimeters that help to decipher retro-tectonic aspects. Along with the geologic sites, there are several physical-geographic sites, such as The Licaş Swallow Hole, The Melekviz intermittent sources and the Izvorul Rece site and there are also archaeological objectives, such as with the Bardos-Bicaz Gorges. The vegetal sector is very well represented, through monuments of nature (*Taxus baccata, Gentiana lueta, Leontopodium alpinum, Cypripedum calceolus*); rarities (*Juniperus sabina, Nigritella rubra, Daphne cneorum, Larix decidua* ssp. *carpatica*) or endemisms (*Aconitum toxicum, Campanula rotundifolia* ssp. *calcicolum, Centaurea melanocalathia*).

The inclusion of Romania into the European Union led to an increase of natural reservations and to more and more numerous propositions for the protection of new regions. After 1989, the Danube Delta was declared a Biosphere Reservation, thus becoming the only morpho-hydrographic unit of this type in the world. In the same period, many of the territories that presented scientific and cultural interests acquired the status of National and Natural Parks. In Romania, 29 such National and Natural Parks were declared, as follows:

1. Danube Delta Biosphere Reservation;
2. Domogled-Valea Cernei National Park;

G. Romanescu et al., *Limnology of the Red Lake, Romania*,
DOI: 10.1007/978-94-007-6757-7_3,

3. Ceahlău National Park;
4. Cozia National Park;
5. Jiu Gorge National Park;
6. Buila-Vânturariţa National Park;
7. Semenic-Caraş Gorges National Park;
8. Retezat National Park;
9. Bicaz Gorges-Hăghimaş National Park;
10. Cheile Nerei-Beuşniţa National Park;
11. Măcinului Mountains National Park;
12. Piatra Craiului National Park;
13. Călimani National Park;
14. Rodnei Mountains National Park;
15. Geopark of the Dinosaurs Ţara Haţegului;
16. Grădiştea Muncelului Natural Park;
17. Iron Gates Natural Park;
18. Vânători-Neamţ Natural Park;
19. Apuseni Natural Park;
20. Mehedinţi Plateau Geopark;
21. Bucegi Natural Park;
22. Comana Natural Park;
23. Maramureşului Mountains Natural Park;
24. Putna-Vrancea Natural Park;
25. Balta Mică a Brăilei Natural Park;
26. Mureş Floodplain Natural Park;
27. Low floodplain of lower Prut;
28. Gorge of upper Mureş;
29. Cefa Natural Park.

The Red Lake is part of the Bicaz Gorges-Hăghimaş National Park.

The Order of the Ministry of Water, Forests, and Environmental Protection no. 7/27.01.1990 and Law no. 5/2000 on the plan of development for the national territory, Section III, protected areas, confirms the foundation of the Bicaz Gorges-Hăghimaş National Park. In this sense it must be mentioned Government Decision no. 230/2003 regarding the delimitation of reservations belonging to the biosphere, national and natural parks.

As a natural protected area, the Bicaz Gorges-Hăghimaş National Park is a point of interest nationwide and it is included in the second category stipulated by the International Union for Conservation of Nature (IUCN) (Grasu et al. 2011). It is situated in the Group of Giurgeului and Hăghimaş Mountains (Curmăturii or Hăşmaş) within the Moldavian-Transylvanian Carpathians (Romanian Carpathians and Transylvanian Depression 1987). The Curmăturii Mountains are also known as the Hăghimaş or Hăşmaş Mountains, which led to their generalized denomination within the specialized literature. In this case, the term Hăghimaş was preferred as it is better known and often used by locals.

It must be mentioned that the delimitation of mountain units has been an intensely disputed subject within the past few years, the reason for which the limits of the Hăghimaş Mountains or even its name are still being questioned. In most specialized works there have been contradicting struggles regarding the delimitation and denomination for the Group of Giurgeului and Hăghimaş Mountains within the Moldavian-Transylvanian Carpathians. The Hăghimaş Peak and the Tulgheş-Valea Rece Peak are considered subdivisions of the Hăghimaş Massif (Mihăilescu 1963). The mountain space between the Bistricioara River (north) and the Trotuş River (south) is divided into the Hăghimaş Mountains, Suhard Mountains, Tulgheş Mountains, Dămuc Mountains and Aramei Mountains (Bojoi 1982 cited by Niculescu and Oancea 1987).

The Bicaz Gorges-Hăghimaş National Park consists mostly in the former Complex Reservation Bicaz Gorges—Red Lake. It is situated entirely in the Hăghimaş Mountains and the Red Lake is situated at the foot of the Suhard and Ucigaşu Mountains. The Hăghimaş Mountains overlap the southern sector of the external Marginea Syncline, delimited by a range of volcanic mountains to the west and by a range of flysch to the east (Mihăilescu 1963; Niculescu and Oancea 1987).

The Hăghimaş Mountains are delimited by the Bistricioara River to the north and by Mihăileni-Valea Frumoasei to the south. The maximum length between the two extremes is 50 km (Grasu et al. 2011).

From a structural-geologic perspective, the Hăghimaş Mountains are part of the Crystalline-Mesozoic compartment of the Moldavian or Tisa-Ciuc Compartment (Grasu et al. 1995, 2010, 2011; Săndulescu 1984). The pre-Alpine crystalline basement and the Mesozoic sedimentary cover engage as pre-Alpine and Alpine overthrust nappes, mostly covered by old post-tectonic deposits or by Neogene vulcanites. The Mesozoic sediments are still conserved on the eastern edge of the crystalline, within the "External Marginal Syncline" (Grasu et al. 2011).

A sector of the Bistriţa Mountains suffers axial elevation and it separates the area into two distinct zones: the Rarău Syncline to the north and the Hăghimaş Syncline to the south. In this case, the Hăghimaş Mountains overlap, for the most part, the Mesozoic syncline on the Bistricioara valley to the north and Frumoasa-Ciuc to the south. The eastern limit with the flysch corresponds to the Pintec, Jidanului, Dămuc, Valea Rece creeks and to the Trotuş sources (Grasu et al. 2011). The Hăghimaş syncline comprises the following structural units, with an underground-surface direction: the sub-Bukovinian Nappe and the Bukovinian Nappe.

The Bicaz Gorges-Hăghimaş National Park is situated on the territory of two counties: Harghita and Neamţ. On the north–south direction, it is approximately 20 km long, while the maximum width is 10 km between the sources of the Licaş creek and the mouth of the Şugău creek. It has a surface of 6,793.64 ha, of which 3,649 ha (54 %) belong to the Neamţ County and 3,144 ha (46 %) to the Harghita County. The northern sector belongs to the area occupied by the Red Lake and Bicaz Gorges and the central and southern areas correspond to the massifs of Ghilcoş, Hăghimaşul Negru, Telecul Mare, Telecul Mic, Hăghimaşul Mare and Fratele. The entire surface of the Bicaz Gorges-Hăghimaş National Park belongs to the calcareous massifs within the main body of the Transylvanian Hăghimaş Nappe (Grasu et al. 2011).

The Bicaz Gorges-Hăghimaş National Park has objectives related to hydrology, hydrogeology, flora, fauna, archaeology and to the geologic substrate. Besides the lake basin being occupied by the Red Lake and by karstic forms dominated by the Bicaz Gorges, there are also numerous geologic sites such as the fossil-bearing points with a biostratigraphic and paleontologic value, a series of petrographic outcrops with paleogeographic significance and some structural perimeters that help to decipher retro-tectonic aspects.

3.1 Geologic Sites

1. The Werfen formation at Poiana Albă (facial and faunistic importance);
2. The Adneth Lyassic at Curmătura (an enclave caught in the mass of the Cretaceous wildflysch, similar to the Adneth Lyassic within the Rarău and Perşani Mountains);
3. The formation with *Acanthicum* at Ghilcoş and Ciofronca (33 types of described ammonites have the halotypes of Ghilcoş and Ciofronca, while some of them became genotypes; halotypes for a species of brachiopods, for 3 bivalves, for an echinus and for 4 algae and foraminifers; the presence of ammonites made the Ghilcoş sequence be proposed as an hypostratotype for the Kimmeridgian-Tithonian limit; it is a typical tanatocenosis) (Grigore 2002);
4. The Kimmeridgian at the origin of the Fagul Oltului creek (lumachelle fossil-bearing nest, a rare case for the Mesozoic of Eastern Carpathians);
5. The fauna of the Stramberg limestones on the Oaia creek (the fossil-bearing points in the Hăghimaş Mountains, mostly those in the Oaia creek, which are third in the world after the fossil-bearing nests of the same age at Wozniki and Kruhel within the Northern Carpathians);
6. The Neocomian of the Ghilcoş Mountain and of the Hăghimaşul Negru Mountains (numerous species of algae and foraminifers that are new for science, with halotypes in the massifs of Ghilcoş, Surduc, Cupaş and Suhărzel);
7. The sites with Urgonian fauna (the species *Cayeuxia piae* and *Cayeuxia moldavica*, described in the Bardos Massif entered the world through paleontologic literature) (Frollo 1940);
8. The fossil-bearing Lyassic and Doggerian on the Ghilcoş creek (the fossil-bearing sequence on the Ghilcoş creek is extremely instructive and one of the few of the Hăghimaş synclines where the intervention of the Eocimerian and intra-Lyassic tectogeneses of the Donetz phase is examined, which affected the Bukovinian sedimentary field);

9. The basalts and serpentinites in the area of the former Bicaz Gorges cabin (the basic magmatites have a special significance from a geo-tectonic and paleogeographic perspective; it is the biggest outcrop in the Hăghimaş Syncline and the only one in the Eastern Carpathians where one can examine the pillow-lava structures occurring today in oceanic rifts) (Grinea 1998);
10. The tectonic windows of the Bicaz Holes and Cheile Mari (sites with a structural-tectonic character that have scientific-instructive importance).

3.2 Physical-Geographic Sites

1. The Licaş Swallow Hole (the most important swallow holes in the Hăghimaş Mountains, known as the Licaş Hole; caving-in swallow hole accessible only to speleologists) (Cristea 1978);
2. The Melekviz intermittent sources (the largest karstic insurgence source in the Hăghimaş Mountains; it is intermittent; around 13,000 l/min discharge; it does not freeze in the winter–Melekvitz means "hot");
3. Izvorul Rece (the second largest karstic source within the Hăghimaş Mountains; resurgence with an intermittent character; discharge of about 9,000 l/min) (Dragomir 2002);

3.3 Exemplary Objectives

The Red Lake (one of the most long-lasting natural dam lakes in Romania, formed in 1837), is best known from this perspective; its name comes from the red color of the water, a reflection of the red color existent in the limestones of the left slop; it has a maximum depth of 10.5 m).

The Bicaz Gorges (gorges with a total length of 8 km: 4 km of limestone in the sector of the Big Gorges and 4 km of Bârnaru conglomerates in the sector of the Small Gorges; there are altitude differences of 350 m, with abrupt walls, extremely appreciated by alpinists). They are the best known and the most visited gorges in Romania.

3.4 Archaeological Objectives

The archaeological site of Bardos-Bicaz Gorges (Swiderian settlement; it belongs to the Epipaleolithic at the end of the Würm glaciation; the site is similar to that of Scaune-Ceahlău and the two are the only remains of the Swiderian-Mazowian culture in Romania, originating in Poland, 20 km north-east from Warsaw) (Bitiri and Căpitanu, 1967).

3.5 Protection of Vegetal Special Groups

Because of altitudinal layering and therefore the diversified climate of the geologic substrate as a petrographic mosaic and considering the different slope orientation, a rich vegetal layer covers the Hăghimaş Mountains. Some species—such as flora rarities or endemisms—are protected by law or proposed for such protection (Grasu et al. 2010, 2012).

Most rarities in the flora and phytocenology are characteristic to the mountain peaks and to the Bicaz Gorges (Horeanu 1979a, b; Mititelu and Nechita 1992, 1993). The massif holds 90 rare, endemic taxons. Negative human intervention determined the authorities to turn the zone into a National Park.

Monuments of nature: *Taxus baccata, Gentiana lueta, Leontopodium alpinum, Cypripedum calceolus.*

Rarities: *Juniperus sabina, Nigritella rubra, Daphne cneorum, Larix decidua* ssp. *carpatica.*

Endemisms:

- General Carpathian endemisms (38 % of the endemisms): *Aconitum toxicum, Campanula rotundifolia* ssp. *calcicolum, Centaurea melanocalathia, Cerastium arvense* ssp. *calcicolum, Cerastium lerchenfeldianum, Cardamine glanduligera, Festuca carpatica, Melampyrum sylvaticum* ssp. *saxosum, Silene zawadzkii, Oxytropis carpatica, Phyteuma vagneri, Pulmonaria rubra* ssp. *filarszkyana, Ranunculus carpaticus, Sesleria heuffleriana, Symphytum cordatum, Thlaspi dacicum, Thymus pulcherrimus, Trisetum macrotrichum.*
- Endemisms for the Romanian Carpathians (26 % of the endemisms): *Aconitum lasianthum, Aconitum moldavicum, Aquilegia nigricans* ssp. *subscaposa, Campanula carpatica, Centaurea triumfetti* ssp. *pinnatifida, Dianthus kitaibelli* ssp. *spiculifolius, Helictotrichon decorum, Hepatica transsilvanica, Sesleria haynaldiana, Silene dubia, Thymus comosus, Thymus bihorensis.*
- Endemisms for the Eastern and Southern Carpathians (18 % of the endemisms): *Achilea schurii, Androsace villosa* ssp. *arachnoidea, Dianthus tenuifolius, Eritrichium nanum* ssp. *jankae, Erysimum witmannii* ssp. *transsilvanicum, Genista tinctoria* ssp. *oligosperma, Gentiana phlogifolia, Hesperis matronalis* ssp. *moniliformis, Scabiosa lucida* ssp. *barbata.*
- Endemisms for the Eastern Carpathians (8 % of the endemisms): *Centaurea carpatica, Hieracium racemosum* ssp. *rehmanni, Primula leucophylla.*
- Endemisms with a limited diffusion in counties or regions (6 % of the endemisms): *Aconitum firmum* ssp. *romanicum, Astragallus römeri, Viola jooi.*
- Local endemisms (4 % of the endemisms): *Astragalus pseudopurpureus, Festuca scoparia* ssp. *lueta.*

Dacian-Balkan elements (around 45 species): *Achillea critmifolia, Alyssum repens, Asperula capitata, Campanula abietina, Centaurea atropurpurea, Cirsium furiens, Crocus heuffelianus, Doronicum carpaticum, Festuca cinerea* ssp. *pallens, Festuca drymeia, Festuca porcii, Helleborus purpurascens, Hieracium* *transsilvanicum, Juncus thomassi, Saxifraga luteo-viridis, Scrophularia laciniata, Sesleria haynaldiana, Silene heuffelli, Taraxacum nigricans, Verbascum lanatum, Viola dacica, Viola declinata.*

Relicts:

- Tertiary relicts: *Erysimum wittmannii, Melampyrum sylvaticum* ssp. *saxosum, Sesleria coerulans, Silene zawadzkii.*
- Glacier relicts: *Arabis alpina, Carex appropinquata, Carex elongata, Carex sempervirens, Leontopodium alpinum, Pinus sylvestris, Poa alpina, Ribes nigrum, Viola biflora.*

Endemic vegetal associations: *(Artemisio)–Gypsophiletum petreae, Sileneto zawadzkii–Saxifrageto luteo–viridis, Cetrario–Vaccinietum gaultherioides, Seslerietum bielzii, Festucetum amethystinae, Seslerietum heufflerianae, Campanula carpaticae–Poëtum rehmannii, Avenastretum decori, Thymo comosi–Caricetum humilis, Juniperetum sabinae, Vaccinio–Pinetum silvestris, Diantheto compacti–Festucetum porcii, Caricetum appropinquatae.*

The list of the protected plants within the Bicaz Gorges-Hăghimaş National Park is remarkably rich (Table 3.1).

3.6 Protection of the Fauna

The fauna in the Hăghimaş Mountains is highly diversified, despite the reduced size of this area (Munteanu 1998). Among the species and genera that are remarkable through their scientific importance (and as such the law protects them) the following species must be mentioned:

- *Odonata* order, with the boreal-alpine species *Coenagrion hastulatum* (CHARP) and the high-altitude species *Aeschna juncea* L. and *Aeschna grandis* L.;
- *Plecoptera* order, with the species *Nemoura fusca* KIS., *Protonemura aestiva* KIS. and *Cloroperla kisi* ZWICK.;
- *Neuroptera* order, with the species *Wesmailius rava* (WITHYT);
- *Heteroptera* order, with the species *Sciocoris umbrinus* (WIFF.);
- *Coleoptera* order, with eight species of *Carabus*, endemic for Romania;
- *Lepidoptera* order, with the Carpathian endemic species *Coscheltellus carna transsylvanica* DANIEL, *Stenoptilia pneumonanthes* SCHL., *Datascia dilucidaria carpathica* SOFFNER, *Parasenia plantaginis carpathica* DANIEL, *Parnassius apollo transsylvanicus* SCHW., *Erebia medusa slovakiana* WARR. and *Erebia melas carpathicola* POPESCU GORJ et ALEX (new species for science);

Hymenoptera order, *Ichneumonidae* Family, with the species *Glypta kowalli* OZOLS (new for science) and *Ichneumon factor* D.T., *Cosmoconus nigriventris* KASP., *Polyblasstus carbonator* KASP., *Tryphon abditus* KASP., *Diblastomorpha*

Table 3.1 List of the protected plants within the Bicaz Gorges-Hăghimaş National Park (after the management plan cited by Grasu et al. 2010)

No.	Species/subspecies	No.	Species/subspecies
1	*Lycopodium clavatum*	58	*Linum extraaxilare*
2	*Lycopodium annotinum*	59	*Koeleria macrantha* ssp. *transsilvanica*
3	*Lycopodium complanatum*	60	*Listera cordata*
4	*Huperzia selago*	61	*Larix decidua* ssp. *polonica*
5	*Achillea oxyloba* ssp. *schurii*	62	*Listeria ovata*
6	*Androsace chamaejasme*	63	*Melampyrum saxosum*
7	*Androsace viillosa* ssp. *arachnoidea*	64	*Microstylis monophyllos*
8	*Aquilegia nigricans*	65	*Minuartia laricifolia* ssp. *kitaibelii*
9	*Arnica montana*	66	*Nigritella rubra*
10	*Astragalus pseudopurpureus*	67	*Orchis coriophora*
11	*Astragalus roemeri*	68	*Orchis militaris*
12	*Campanula carpatica*	69	*Orchis morio*
13	*Centaurea atropurpurea*	70	*Orchis tridentata*
14	*Centaurea kotschyana*	71	*Orchis ustulata*
15	*Centaurea pinnatifida*	72	*Oxytropis carpatica*
16	*Cephalanthera damasonium*	73	*Oxytropis halleri*
17	*Cephalanthera rubra*	74	*Phyteuma confusum*
18	*Cerastium transsilvaticum*	75	*Phyteuma tetramerum*
19	*Cerastium arvense* ssp. *lerchenfeldianum*	76	*Phyteuma vagneri*
20	*Cirsium decussatum*	77	*Pinguiculla vulgaris*
21	*Cerastium fontanum* ssp. *macrocarpum*	78	*Pinus silvestris*
22	*Cirsium furiens*	79	*Poa molinerii*
23	*Crepis conyzifolia*	80	*Poa rehmanii*
24	*Crepis jackuinii*	81	*Poa stiriaca*
25	*Cypripedium calceolus*	82	*Potentilla rupestric*
26	*Dactylorhiza cordigera*	83	*Primaula elatior* ssp. *leucophylla*
27	*Dactylorhiza incarnata*	84	*Primula farinosa*
28	*Dactylorhiza maculata*	85	*Primula halleri*
29	*Dactylorhiza sambucina*	86	*Ranunculus carpaticus*
30	*Daphne cneorum*	87	*Ribex nigrum*
31	*Dianthus petraeus* ssp. *spiculifolius*	88	*Saussurea alpina*
32	*Dianthus tenuifolius*	89	*Scabiosa columbaria* ssp. *pseudobanatica*
33	*Doronicum carpaticum*	90	*Seseli gracile*
34	*Draba kotskyi*	91	*Scabiosa lucida* ssp. *barbata*
35	*Eritrichium nanum* ssp. *jankae*	92	*Sempervivum montanum*
36	*Erysimum wittmanni*	93	*Scorzonera purpurea* ssp. *rosea*
37	*Festuca carpatica*	94	*Seseli rigidum*
38	*Fastuca vesicolor*	95	*Scrophularia heterophylla* ssp. *laciniata*
39	*Galanthus nivalis*	96	*Sesleraia bielzii*
40	*Galium pumilum*	97	*Sedum telephium* ssp. *fabaria*
41	*Gentiana crucisata* ssp. *phlogifolian*	98	*Semeprvivum marmoreum*
42	*Gentiana lutea*	99	*Silene nutans* ssp. *dubia*

(continued)

Table 3.1 (continued)

No.	Species/subspecies	No.	Species/subspecies
43	*Gentiana orbicularis*	100	*Silene zawadskii*
44	*Gentiana pneumonanthe*	101	*Soldanella montana*
45	*Goodyera repens*	102	*Taxus baccata*
46	*Gymnadenia conopsea*	103	*Thalspi dacicum*
47	*Gypsophila petraea*	104	*Thymus bihorensis*
48	*Helictotrichon decorum*	105	*Thymus comosus*
49	*Hepatica transsilvanica*	106	*Traunsteinera globosa*
50	*Hieracium pojoritense*	107	*Trisetum alpestre*
51	*Hesperis matronalis* ssp. *cladotrichus*	108	*Trollius europaeus*
52	*Jovibarba heuffeli*	109	*Vaccinium gaultheroides*
53	*Iris aphylla* ssp. *hungarica*	110	*Viola alpina*
54	*Jovibarba sobolifera*	111	*Viola adacica*
55	*Juniperus sabina*	112	*Viola jooi*
56	*Leontopodium alpinum*	113	*Waldsteinia geoides*
57	*Jurinea mollis* ssp. *transsilvanica*	114	*Waldsteinia ternata*

bicornis (BOIE), *Calajoppa lapidator* (F.), *Poemenia collaris* HAUPT., *Megarhyssa emarginatoria* (THUNB.) (Grasu et al. 2010).

In the Hăghimaş Mountains, there are 82 species of animals protected by law (Table 3.2). Some of them are new for science, while some others are endemic or very rare.

3.7 Sites Proposed to be Included in the Perimeter of the Bicaz Gorges-Hăghimaş National Park

1. Munticelu-Criminiş area (Grasu et al. 2008, 2011):
    (a) Munticelu Massif (the only spot in the Hăghimaş Mountains where the abnormal position of the limestone mass within the Transylvanian nappe is clearly visible (Uhlig 1907); the limestone massifs within the Hăghimaş Mountains appear as overthrust nappes, thus forming large covering fragments);
    (b) Şugău Gorges (the lateral marmites situated at 30 m high and the existence of riverbed marmites support the epigenetic hypothesis of the gorges; the existence of an accumulation of travertine, with a current formation, as a unique place in Romania);
    (c) The Munticelu cave (this case has an authentic potholing dowry; it was discovered in 1973; remains of *Ursus spelaeus*; the highest density of candle-shaped stalagmites in Romania (Acrâşmăriţei and Ciobanu 1973; Giurgiu 1990; Grasu et al. 2011); the existence of fistula stalagmites);

Table 3.2 Animal species protected by law in the Bicaz Gorges-Hăghimaş National Park

No.	Species	OUG act 57/20.06.2007
Invertebrates		
1	*Austropotamobius torrentium*	+
2	*Erebia medusa slovakiana WARR*	+
3	*Erebia melas carpathicola POPESCU GORJ*	+
4	*Parassius apollo transsylvanicus SCHW*	+
Fish		
5	*Barbus meridionalis*	+
6	*Cottus gobio*	+
7	*Thymallus thymallus*	+
Amphibians		
8	*Bombina variegata*	+
9	*Bufo bufo*	+
10	*Bufo viridis*	+
11	*Rana temporaria*	+
12	*Salamandra salamandra*	+
13	*Triturus alpestris*	+
14	*Triturus cristatus*	+
15	*Triturus montandoni*	+
16	*Triturus vulgaris*	+
Reptiles		
17	*Anguis fragilis*	+
18	*Coronella austrica*	+
19	*Larceta muralis*	+
20	*Vipera berus*	+
Birds		
21	*Alcedo athis*	+
22	*Anas platyrinchos*	+
23	*Apus apus*	+
24	*Aquila Chrysaetos*	+
25	*Aquila pomarina*	+
26	*Athene noctua*	+
27	*Aythya nyroca*	+
28	*Botaurus stellaris*	+
29	*Bubo bubo*	+
30	*Calandrella brachydactyla*	+
31	*Caprimulgfus europaeus*	+
32	*Chlidonias niger*	+
33	*Ciconia ciconia*	+
34	*Cinclus cinclus*	+
35	*Circaetus gallicus*	+
36	*Corvus corax*	+
37	*Coracias garrulus*	+
38	*Cygnus cygnus*	+

(continued)

Table 3.2 (continued)

No.	Species	OUG act 57/20.06.2007
39	*Dendrocopus leucotos*	+
40	*Dendrocopus medius*	+
41	*Dendrocopus syriacus*	+
42	*Dryocopus martius*	+
43	*Emberiza cia*	+
44	*Emberiza hortulana*	+
45	*Erithacus rubecula*	+
46	*Falco naumanni*	+
47	*Falco peregrinus*	+
48	*Ficedula parva*	+
49	*Gavia artica*	+
50	*Jynx torquilla*	+
51	*Lanius minor*	+
52	*Lanius collurio*	+
53	*Larus minutus*	+
54	*Lullula arborea*	+
55	*Mergus albellus*	+
56	*Monticola saxatilis*	+
57	*Motacilla alba*	+
58	*Motacilla cinerea*	+
59	*Nucifraga caryocatactes*	+
60	*Picoides tridactylus*	+
61	*Picus canus*	+
62	*Podiceps auritus*	+
63	*Regulus regulus*	+
64	*Regulus ignicapillus*	+
65	*Sitta europaea*	+
66	*Sterna hirundo*	+
67	*Streptopelia turtur*	+
68	*Tetrao urogallus*	+
69	*Tetrastes bonasia*	+
70	*Tichodroma muraria*	+
71	*Troglodytes troglodytes*	+
72	*Tringa glareola*	+
73	*Turdus philomelos*	+
74	*Turdus viscivorus*	+
Mammals		
75	*Barbastella barbastellus*	+
76	*Canis lupus*	+
77	*Felis silvestris*	+
78	*Lynx lynx*	+
79	*Martes martes*	+
80	*Myotis myotis*	+
81	*Myotis dascymene*	+
82	*Rupicapra rupicapra*	+

(d) The formation with *Aptychus* on the Criminiş Mountain (the only spot in the Hăghimaş Syncline where the Malm of the Bukovinian autochthon, in pelagic facies of the Formation with *Aptychus*, may constitute an olistolith body);
(e) The Serpentinites of Crucea lui Ţepeş (the emergence of ultrabasic rocks within the ophiolite complex specific to the oceanic crust within the Transylvanian "fosse" on which the Malm-Urgonian calcareous series in the Hăghimaş Nappe accumulated).

2. The sub-Bukovinian leveling nappe outlier in the area of the Gura Dămuc-Ciofu creek (leveling nappe outlier; a lithostratigraphic sequence of the sub-Bukovinian unit, of which the Doggerian provided unique fauna for the Eastern Carpathians (Grasu, Turculeţ, 1965); paleogeographic importance);
3. The klippes incorporated into the wildflysch formation of Piatra Luciului (neritic Lyassic and Doggerian, similar to the Bukovinian series within the rest of the syncline, "plucked out" of the autochthon's succession and included in the wildflysch deposits);
4. The formation with *Aptychus* in the Dămuc basin (fossil Bukovinian autochthon with aptychi and ammonites); given the Malm-Neocomian age, it represents the stratigraphic equivalent of the Malm-Urgonian in the Hăghimaş Transylvanian Nappe;
5. The Hăghimaş breccia (the deposits at the beginning of the alpine cycle; coluvial-diluvial formation accumulated before the Seisian transgression, probably during the Permian, in the rifting phase of the Carpathian basin);
6. The autochthonous basic magmatites (autochthonous basic magmatites within the Crystalline-Mesozoic area of the Eastern Carpathians; the two types of magmatites are considered products of the processes associated to the opening of the External Dacides rift).

References

Acrâşmăriţei N, Ciobanu M (1973) Peştera Munticelu. Studii şi Cercetări Ştiinţifice, Muzeul de Ştiinţe Naturale Piatra Neamţ, Geologie, Piatra Neamţ 2

Bitiri M, Căpitanu V (1967) O nouă aşezare swideriană în Carpaţii Orientali. Stud Res Ancient Hist Bucharest, vol 18(1)

Cristea E (1978) Munţii Hăşmaş şi Staţiunea Lacu Roşu. Sport-Turism Publishing House, Bucharest

Dragomir GP (2002) Ape plate din izvoare carstice în Munţii Hăghimaş. EcoCarst J, Bucharest 3

Frollo M (1940) Etude pétrographique des séries sédimentaires de la vallée du Bicaz–Carpathes Orientales. These, Paris

Giurgiu M (1990) Peşterile din Masivul Hăşmaş. Buletin CSER, Baraolt

Grasu C, Turculeţ I (1965) Asupra vârstei recifului de la Gura Dămucului. SSNG Geology Communications, III, Bucharest

Grasu C, Turculeţ I, Catana C, Niţă M (1995) Petrografia Mezozoicului din "Sinclinalul Marginal Extern". Romanian Academy Publishing House, Bucharest

Grasu C, Brânzilă M, Micläuş C, Baciu S (2008) Geological evidence supporting the proposal of extension of the Bicaz Gorge-Hăşmaş Mountains National Park, vol 75. Annals of the Romanian Geologic Institute, Bucharest (Special Issue)

Grasu C, Micläuş C, Brânzilă M, Baciu DS (2010) Munţii Hăşmaşului. Monografie geologic şi fizico-geografică. "Al.I.Cuza University" Publishing House, Iaşi

Grasu C, Brânzilă M, Micläuş C, Baciu DS (2011) Parcul Naţional Cheile Bicazului-Hăşmaş. Ghidul siturilor geologice şi fizico-geografice. "Al.I. Cuza" University Publishing House Iaşi

Grasu C, Micläuş C, Brânzilă M, Baciu DS (2012) Sinclinalul Mezozoic Tulgheş-Hăşmaş-Ciuc. "Al.I. Cuza" University Publishing House Iaşi

Grigore D (2002) Formaţiunea cu Acanthicum din regiunea Lacu Roşu (Masivul Hăghimaş-Carpaţii Orientali), posibil hipostratotip al limitei Kimmeridgian-Tithonic. PhD thesis, Alexandru Ioan Cuza University, Iaşi

Grinea D (1998) Studiul mineralogic şi geochimic al rocilor eruptive bazice mezozoice din Munţii Hăghimaşului şi Munţii Tulgheşului. PhD thesis, Alexandru Ioan Cuza University, Iaşi

Horeanu C (1979a) Flora rezervaţiei naturale Munticelu-Cheile Şugăului (Jud. Neamţ). Annals of the Museum of Natural Sciences Piatra Neamţ, Biology Series, Piatra Neamţ

Horeanu C (1979b) Consideraţii asupra florei şi vegetaţiei complexului carstic Cheile Bicazului-Lacu Roşu. Museum of Natural Sciences Piatra Neamţ (manuscript)

Mihăilescu V (1963) Carpaţii Sud-estici. Scientific Publishing House, Bucharest

Mititelu D, Nechita N (1992) Flora muntelui Hăşmaş şi din Cheile Bicazului. Studies and researches. Museum of Natural Sciences, Piatra Neamţ

Mititelu D, Nechita N (1993) Contribuţii la studiul florei şi vegetaţiei din Masivul Hăşmaş şi Cheile Bicazului. Studies and Researches. Museum of Natural Sciences, Biology, Piatra Neamţ

Munteanu D (1998) Cercetări asupra avifaunei Munţilor Hăghimaş, vol 4. Banat Annals, Sciences of Nature, Timişoara

Niculescu G, Oancea D (1987) Munţii Giurgeului şi Curmăturii. Geografia României, vol III. Romanian Academy Publishing House, Bucharest

Săndulescu M (1984) Geotectonica României. Technical Publishing House, Bucharest

Uhlig V (1907) Uber die Tektonic der Karpathen. Sitzungsber. k. Akad. Wiss, XVI, Vienna

Chapter 4
Paleogeographic Evolution of the Hydrographic Basin and the Lacustrine Basin

Abstract The central-longitudinal area of the Eastern Carpathians belongs to the Crystalline-Mesozoic unit. This area, known as the Carpathians, belongs to the Crystalline-Mesozoic unit, whose sector is known as the Moldavian Compartment (the Tisa-Ciuc Compartment) (Grasu C, Miclăuş C, Brânzilă M, Baciu DS (2010) Munţii Hăşmaşului. Monografie geologic şi fizico-geografică. "Al.I.Cuza University" Publishing House, Iaşi; Mutihac V (1990) Structura geologică a teritoriului României. Technical Publishing House, Bucharest). To the east, the Mesozoic sedimentary is known as the External Marginal Syncline. It is separated into two compartments: the Hăghimaş syncline in the south and Rarău in the north. The Hăghimaş Mountains, which also comprise the Red Lake lacustrine basin, correspond to the Mesozoic syncline, located between Frumoasa-Ciuc in the south and Bistricioara in the north (Grasu C, Miclăuş C, Brânzilă M, Baciu DS (2010) Munţii Hăşmaşului. Monografie geologic şi fizico-geografică. "Al.I.Cuza University" Publishing House, Iaşi). The Red Lake was formed in the summer of 1837, when, after heavy rainfalls, a landslide diluvium fell from the Ghilcoş (Ucigaşu) Mountains and blocked the stream of the Bicaz brook. The Ghilcoş-Hăghimaş-Mezinul Crest is located in the northern compartment of the central unit within the Eastern Carpathians. The most frequent rocks within the Red Lake hydrographic basin are the following: limestones and dolomites (Triassic), limy sandstones (Lias-Dogger), sandstones and conglomerates (Barremian-Albian) and crystalline schists (Maastrichtian). There are five sedimentation cycles: the first comprises sequences of conglomerates, sandstones, loam and limestones; the second is associated with Lias limestones; the third is represented by conglomerates, calcareous sandstones and Dogger-Portlandian grey limestones; the fourth mostly comprises Neocomian whitish grey limestones; and the last one comprises conglomerates (along the Bicăjel valley, at the contact with the slopes) with loam and sandstone intercalations (the bottom of the Bicăjel valley).

The central-longitudinal area of the Eastern Carpathians belongs to the Crystalline-Mesozoic unit, whose sector is known as the Moldavian Compartment (the Tisa-Ciuc Compartment) (Grasu et al. 2010; Mutihac 1990). It comprises a

G. Romanescu et al., *Limnology of the Red Lake, Romania*,
DOI: 10.1007/978-94-007-6757-7_4,

pre-Alpine crystalline basement and a Mesozoic sedimentary cover, structured in pre-Alpine and Alpine in overthrust sheets, covered from place to place by older post-tectonic deposits or by neogene volcanites.

The Mesozoic sedimentary in the east is known as the External Marginal Syncline. Its axial elevation separates it into two compartments: the Hăghimaş syncline in the south and Rarău in the north. The Hăghimaş Mountains, which also comprise the Red Lake lacustrine basin, correspond to the Mesozoic syncline, located between Frumoasa-Ciuc in the south and Bistricioara in the north (Grasu et al. 2010).

The Moldavian Compartment of the Crystalline-Mesozoic unit belongs to the Central Eastern Carpathian nappes, which trigger the crystalline basement. In the Hăghimaş syncline, there are the following structural units: The sub-Bukovinian and Bukovinian Nappes. The sub-Bukovinian nappe appears on limited areas, as glacier-planed spots on top of the Bukovinian Nappe. The Bukovinian Nappe has no crystalline basement and it corresponds to the obducted Transylvanian nappes (the Hăghimaş Nappe).

The Red Lake is the best-known Romanian water body formed after a landslide that blocked the stream of the Bicaz brook. Nonetheless, there has been no complete geographic or limno-ecological study focused on it. From a morphologic and morphographic perspective, there have been several attempts, most of them incomplete. The best-known paper regarding this subject is the one elaborated by Pandi G. in 2004 but it was published in Hungarian and it does not include a modern, interdisciplinary, geo-ecologic view on the lacustrine basin system.

The Red Lake is included in the category of natural dam lakes. It was formed in the summer of 1837, when, after heavy rains, a landslide diluvium fell from the Ghilcoş Mountains (Ucigaşu) and blocked the stream of the Bicaz brook. Behind the landslide wave, a lake appeared and the trunks of the forest moved at the same time with the slope remaining in the lake (Bojoi 1968a, b; Mihăilescu 1940, cited by Popp 1941, Pişota and Năstase 1957; Preda 1967; Grasu and Turculeţ 1998; Grasu et al. 2010).

There are several lakes in the Carpathian Mountains formed by a landslide which blocked a stream (Vulturilor, Crucii (Cuejdul), Bolătău, Dracului, etc.) but the best known and most researched is the Red Lake. It is also famous because it is near a transport system that connects Moldavia to Transylvania. In 1910, the road connecting the Gheorgheni locality to the "Gâtul Iadului" point (confluence with Bicăjel) was inaugurated. The Bicaz Gorges was penetrated only in 1937, when the connection between the two historical regions became complete.

With up to date, precise and modern equipment, as well as with great professionals, a series of measurements were done regarding the morphometry and morphography of the Red Lake. They constitute the basis for complex geographic, geological, biological and economic studies, etc. and they will open the way for new interdisciplinary directions regarding limno-ecological studies.

The surface geomorphologic methodology was tested and applied to the morpho-bathymetry of lacustrine basins. These are the same principles of another field, collateral to limnology but geographically important. Most of the parameters known for the emersed territories are also to be found in a submerged environment, but at another scale and with different influences. The geomorphologic and geologic parameters will contribute to understanding the evolution system of a lacustrine basin formed by blocking a stream.

The most important studies, which dealt with the geologic, geomorphologic and hydrologic features of the Hăghimaş Mountains (Red Lake), were elaborated by the following authors: Bojoi (1968a, b), Cărăuşu et al. (1971a, b), Ciaglic (2005), Dobrescu and Ghenciu (1970), Ghenciu (1968a, b), Ghenciu and Apăvăloaie (1969), Ghenciu and Cărăuşu (1970), Gîştescu (1971), Grasu and Turculeţ (1980), Grasu et al. (2010, 2011, 2012), Herbich (1878), Puşcariu (1939), Mihăilescu (1940), citat de Popp (1941), Pandi and Magyari (2003), Pandi (2004), Pandi and Buzilă (2004), Pelin (1967, 1971), Popp (1941), Senchea (1948), Popescu and Dimitriu (1950), Pişota and Năstase (1957), Preda and Pelin (1963), Preda (1967, 1969, 1971, 1976), Romanescu (2009a, b, c), Udrişte (1963), Xantus and Xantus (1999), etc.

The Ghilcoş-Hăghimaş-Mezinul Crest is located in the northern compartment of the central unit within the Eastern Carpathians. It corresponds to the central part of the Ghilcoş-Hăghimaş-Ciuc marginal lacustrine basin. It is included in the Crystalline-Mesozoic unit (Băncilă 1941, 1958; Ciaglic 2005) (Fig. 4.1).

The peak comprises metamorphic and sedimentary rocks, shaped as parallel ribbons, with a north–southward orientation (Băncilă 1941). They have a three-area distribution: with metamorphic rocks of the mesozone (gneissic diorites); with epizone schists, towards the west (quartz-chlorite schists, calcite with phyllites); with mezo-epizone rocks (garnet mica schists, chlorite schists with biotote) (Ciaglic 2005; Grasu et al. 2010).

Fig. 4.1 The Red lake and the landslide diluvium (*left*) that created a barrage in the stream of the Bicaz brook

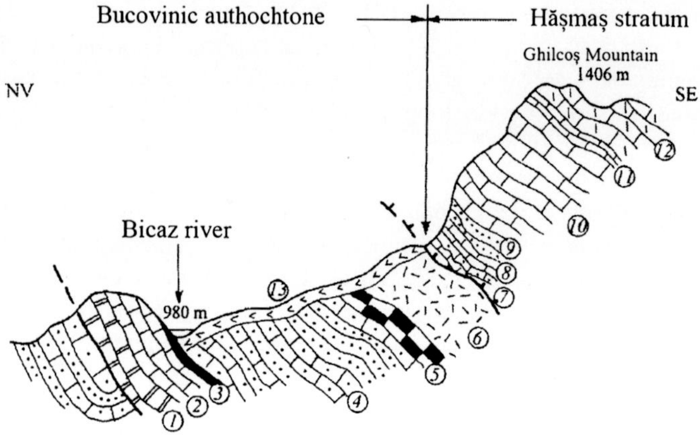

Fig. 4.2 Geologic section in the northern abrupt landform of the Ghilcos mountain (Grasu et al. 2010). The Bukovinian autochthon: *1*.Campilian-Anisian (dolomites); *2*. Middle and Upper Triassic (dolomite and limestones); *3*. Lias Hirlatz; *4*. Dogger; *5*. Callovian-Oxfordian (jaspers and radiolarites); *6*. Barremian-Albian (wildflysch); The Hăghimaş nappe: *7*. Kimmeridgian (nodular-like *red* limestones); *8*. Kimmeridgian (*grey-greenish* limestones); *9*. Kimmeridgian (silty sandy loams); *10*. Tithonian (Stramberg limestones); *11*. Neocomian; *12*. Urgonian; *13*. Landslide body

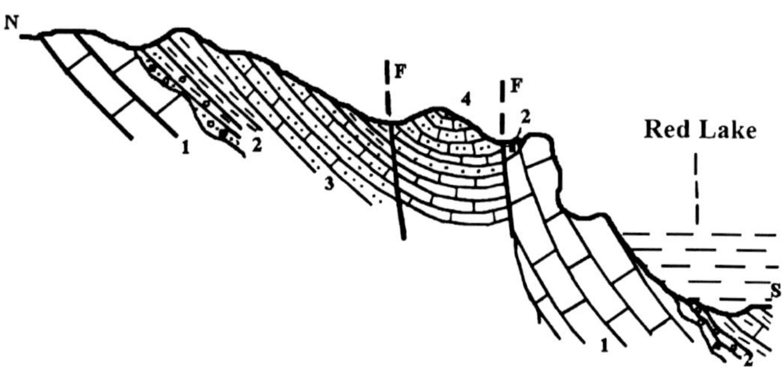

Fig. 4.3 Geologic section of the Mesozoic deposits on the *left* bank of the Red Lake (Preda 1976). *1*. Dolomites and dolomitic limestones—Triassic; *2*. Limestone conglomerates and micaceous sandstones—Domerian-Toarcian; *3*. Gravely limestones and *grey* sandstones—Aalenian-Bajocian; *4*. Gravely, oolitic, and ferruginous limestones—Bajocian-Bathonian-Callovian; *F*—fault

The entire crystalline is extremely curved and fragmentary, with a north–north-westward/south–south-westward general direction of the curves. On top of the Bukovinian autochthon, there is the intensely curved Hăghimaş nappe (Figs. 4.2, 4.3).

The most frequent rocks within the Red Lake hydrographic basin are the following: limestones and dolomites (Triassic), limy sandstones (Lias-Dogger), sandstones and conglomerates (Barremian-Albian) and crystalline schists (Maastrichtian) (Fig. 4.4).

Fig. 4.4 Karst landform of the Hăghimaş mountains

In the north and northeast, there are two sedimentary Mesozoic areas on top of the crystalline. They comprise of Triassic, Jurassic and Cretaceous (Urgonian) limestones, loams, sandstones and Cretaceous conglomerates (Preda and Pelin 1963). The sedimentary deposits are rather thick and they were deposited during five sedimentation cycles, separated by five orogenetic stages (Preda and Pelin 1963).

4.1 The Sedimentation Cycles

Cycle I comprises a sequence of 15–20 m thick Triassic conglomerates; 10 m thick quartz sandstones; thin intercalations of siliceous loam; 200 m thick hard loamy limestones, close to dolomitic limestones; and massive Norian limestones.

Cycle II comprises Lias limestones in Hirlatz facies.

Cycle III comprises conglomerates, calcareous sandstones and Dogger-Portlandian grey limestones. In the Red Lake area, they reach 200 m. On top of them there is a (Calovian-Oxfordian) jasper complex intercalated by clay loams. The presence and position of this complex determine the existence of distinct hydro-geological subunits in karst (Ciaglic 2005).

Cycle IV is formed of Neocomian whitish grey limestones, loamy limestones of Lower Cretaceous, Urgonian limestones and Stramberg limestones (Băncilă 1958). The whole package is 400–500 m thick. The last two occupy the highest part and they represent the most important reservoir of underground waters.

Cycle V comprises conglomerates (along the Bicăjel valley, at the contact with the slopes) with loam and sandstone intercalations (the bottom of the Bicăjel valley). The conglomerates are well developed on the bottom of the Oaia brook—Red Lake Depression, occupying the largest part of the sector downstream from the Red Lake (Ciaglic 2005).

4.2 Geologic Deposits

The Mesozoic deposits are disposed as a large syncline, with a northwestward/southeastward orientation. The syncline sedimentary package was strongly tectonized during the orogenetic phases that occurred between the sedimentation cycles. The tectonic movements led to the appearance of certain faults that segmented the sector of the gorges and the abrupt rocks flanking the crest (Fig. 4.5).

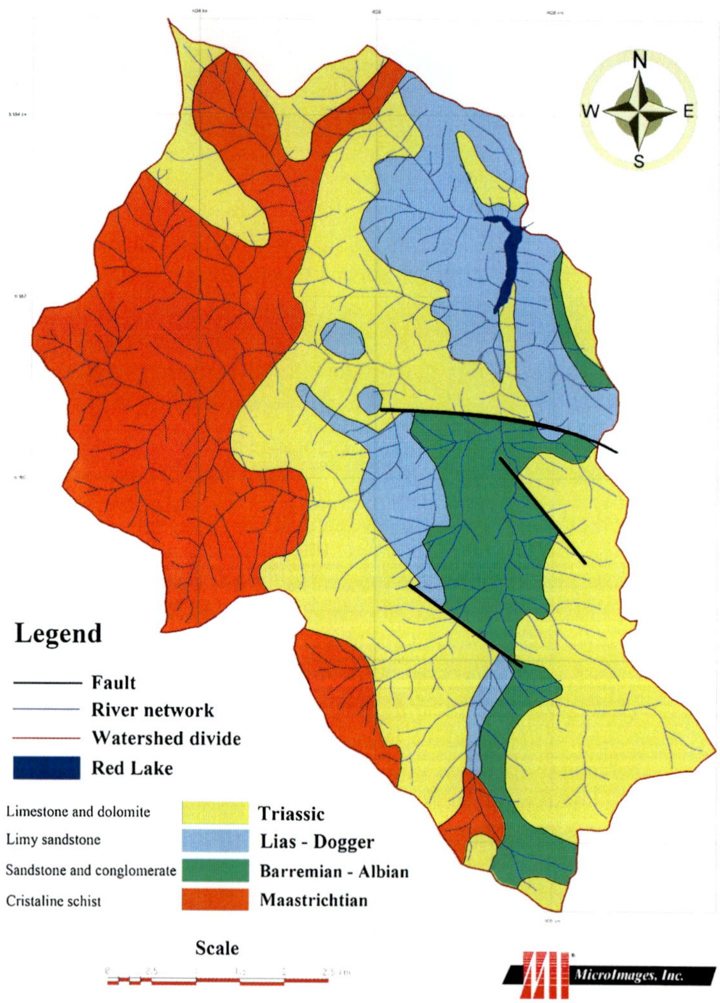

Fig. 4.5 Geologic map of the Red Lake hydrographic basin (Hăghimaş mountains)

During the tectonic movement, the flakes were repositioned on the vertical, independently, elevating Hăghimaşul Mare and Mezinul at the level of the northern extremity and sinking them at the level of the southern extremity (Băncilă 1941). In Hăghimaşul Negru and Ghilcoş, the movements had an inverted orientation. The Mesozoic sedimentary comprises the Bukovinian Autochthon (The sub-Bukovinian Nappe) and the Hăghimaş Nappe (the Hăghimaş Transylvanian Nappe).

The sub-Bukovinian Nappe has the shape of glacier-planned nappe outliers and of tectonic windows. In the Hăghimaşul syncline, it presents itself as a glacier-planned nappe outlier at the confluence between Dămuc and Bicaz, as well as in the tectonic window of Tomeşti (Grasu et al. 2010). This way, there are the following sequence of layers: Triassic (sandstones and Seissian conglomerates, with the dolomitic complex on top); Lias (quartz rubbles, limonitic rubbles, interposed between dolomites and Dogger limestones); Dogger (hard massive limestones, strongly fissured; nodular-like, grey-blackish and fossiliferous limestones; and tabular, black, Muscovite limestones); Tithonian-Neocomian (rudites and polymictic micro-rudites; calcareous litharenites, gravely biopelsparite, etc.).

The Hăghimaş nappe is part of the Transylvanian nappes (Perşani, Olt, and Hăghimaş), with an inside position. In the Hăghimaş syncline, the allochthonous formations, belonging to the Transylvanian series, have the shape of shreds or sedimentary klippes (olistoliths) caught in the mass of the Cretaceous wildflysch deposits (Grasu et al. 2010). The sequence of the following stratigraphic positions can be observed: Triassic (olistolith calcareous blocks in the wildflysch mass); Lias in Adneth facies (red clays and tabular limestones) (Herbich 1878); Dogger (blocks of ferruginous oolitic limestones); Malm (Neojurassic-Urgonian limestones: nodular-like red limestones, grey-green limestones); Neocomian (pelitomorphic limestones with a sub-lithographic aspect, chocolate-yellow, or dark grey); Lower Barremian—Aptian (massive Urgonian limestones—white, grey, yellowish, pink, red, with a pelitomorphic, arenitic or ruditic aspect) (Dragastan 1975; Grasu 1971; Săndulescu 1975).

Fig. 4.6 Ucigaşul peak and the falling cornice of the landslide body

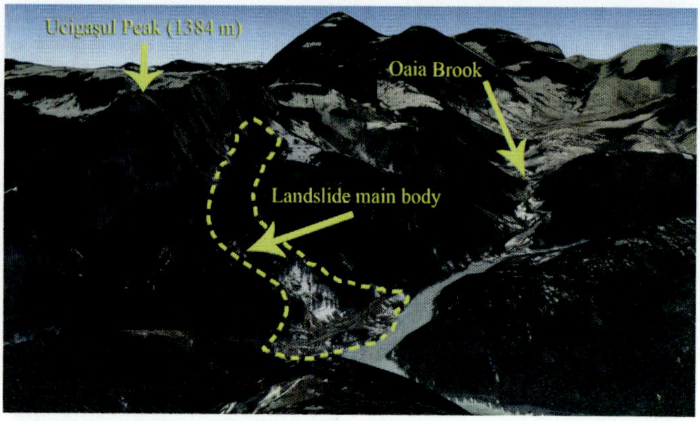

Fig. 4.7 Landslide main body and the appearance of the Red Lake

The landslide mass of the Ghilcoş Mountain (Ucigaşul Peak) (Fig. 4.6) is Cretaceous, wildflysch. It was moved by heavy rains or by an earthquake (Fig. 4.7).

The first pseudoscientific description of the lake was elaborated in 1857 and the first cottage was built as late as 1925. Thus, there are still questions regarding the year and the cause of the landslide.

The landslide affected the entire western slope of the Ghilcoş Mountain (Herbich 1878; Mihăilescu 1940; Zaruba and Mencl 1954) (Fig. 4.8). The landslide comprised three phases: the formation of debris on the northern slope of the Ghilcoş Mountain, during the last glacial era; the fact that the lower part of the debris decomposed, which led to a mix of blocks and clay; the mix becoming wet and causing the landslide, after the heavy rains of 1837 (Mihăilescu 1940). There are also mentions of an earthquake moving the slope diluvium (Gâştescu 1971; Grasu et al. 2010). The dam, which totally closed the stream of Bicaz, is made of

Fig. 4.8 Toe of the landslide and image of the Suhardul Mic peak

Fig. 4.9 Threshold and mouth of the lacustrine basin lacustrine basin

hard, erosion-resistant rocks. The dam has the shape of a threshold, similar to the steep slope, where the excess water is carried (Fig. 4.9).

A determining role in the formation of the clays that contributed to the landslide was played by the formation of *acanthicum* (red limestones, loams, and pelitoidal sandstones) (Preda 1967). The layer sequence appears normal: Triassic dolomites; Hirlatz Lias; Dogger; Callovian-Oxfordian jaspers; formation with *acanthicum*; Stramberg limestones; Neocomian; and reef Urgonian (Grasu et al. 2010). Between the jaspers and formation with *acanthicum* there is the Cretaceous wildflysch (Săndulescu 1975). The Ghilcoş Mountain is a large shred covering the Hăghimaş Nappe. In consequence, the wildflysch deposit formed the landslide bed.

On January 23, 1938, a Moldavian-type earthquake occurred in the Vrancea region, with a 6° magnitude on the Richter scale (8° on the Mercalli scale). The seismic movements could be felt from a significant distance, from Lvov to Sevastopol and Istambul (Atanasiu 1961).

From a tectonic perspective, the region of the Hăghimaş Mountains is included in the ample movements that took over the entire mountain area of Eastern Carpathians. In the Hăghimaş syncline, certain leveling nappe outliers represent the elements of the sub-Bukovinian Nappe (Atanasiu 1928). They represent upside-down sequences of the Crystalline-Mesozoic area (Gura Dămuc, Bâtca Rotundă) pushed over the Sinaia flysch on the alignment of the central-Carpathian fault (Băncilă 1958; Săndulescu 1975).

The sequence comprises deposits belonging to Triassic, Lyassic, Doggerian and Neocomian. The deposits are engaged in a structure of small nappes separated by a fault, leaning against the overthrust plane (Grasu and Turculeţ 1965; Grasu 1976; Săndulescu 1975). The nappe outlier overlaps the Sinaia Formation within the Ceahlău Nappe.

The process of small nappe formation took place in the Meso-Cretaceous tectogeneses. The internal structure in the sub-Bukovinian field took place in the Austrian-Alpine tectogeneses. The moment the wildflysch formation accumulated in the Bukovinian Nappe, the sub-Bukovinian field was already covered by seawater (Grasu et al. 2010, 2012).

The structural lithostratigraphic diversity above the leveling nappe outliers and lamellas within the sub-Bukovinian Nappe is the consequence of various factors. Among them, there are the wrinkle-shaped appearance of the sub-Bukovinian field (truncated stratigraphic columns, determined by interruptions of the sedimentation process); the accentuated dismantling of the sequences during the Meso-Cretaceous overthrust process of the Bukovinian Nappe, etc. (Bojoi 2000; Grasu et al. 2010, 2012).

Most of the Hăghimaş syncline belongs to the Bukovinian Nappe. This basement nappe also engages re-Alpine crystalophyllic formations. Its overthrust plane intersects different-age formations (crystalline schists, Triassic deposits, the Formation with *Apthycus*, Malm-Neocomian deposits), etc.

Over the Meso-Cretaceous tectogeneses older deformations intervened; they have low amplitude and they are intra-Triassic, intra-Lyassic and post-Bathonian. They are marked by an angular lack of concordance (Săndulescu 1975; Grasu et al. 2010, 2012).

The first lack of discordance appears in the Werfenian (Triassic) base. The second lack of discordance occurs in the post-Anisian. In the Donetz phase within the intra-Lyassic—unfolded after the accumulation of the Hierlatz limestones –, a series of folds with a north-south orientation and inclined flanks appeared (40–50°) (Săndulescu 1975). The interruption, followed by folding, is explained through the discordant character of the Doggerian. The syncline with crystalline schists in the west of Suhardul Mare belongs to the Donetz, the anticline structure at the sources of the Oaia brook and the anticline at the sources of the Fagului brook (the Doggerian deposits repose discordantly on the Triassic dolomites or directly on the crystalline).

The Bukovinian Nappe also presents a series of structures belonging to the Nevadian and to the Austrian-Alpine tectogeneses. In the Austrian-Alpine phase, the frontal anticline of the Bukovinian Nappe (the Dămuc "crest" and the Lunca anticline) is outlined. The large structures of the Bukovinian Nappe have a regional character and they are related to the mid-Cretaceous deformation phase. The Dămuc anticline emerged during the pre-paroxysmal phases, being folded and pushed over the flysch in the Austrian deformations and resumed in the late tectogeneses in the Miocene. South from the Pavaş Mountain, the Dămuc anticline disappears under the Formation with *Apthycus* and then it continues with the Lunca anticline (axis pegged out by dolomitic buttonholes, dolomites with Doggerian accents, Crystalline and Triassic dolomites) (Bojoi 2000; Grasu et al. 2010).

The Dămuc-Lunca "crest" anticline is flanked westwards by the wide riverbed of a syncline where the wildflysch incorporates the Hăghimaş Nappe (Săndulescu 1975). North from the Bicaz Valley, it presents two small synclines with a cover that is post-tectonic, ruditic and with Bârnadu Conglomerates.

The main deformation of the Bukovinian Nappe occurred after the accumulation of the wildflysch and before the deposition of the Vraconian-Cenomanian Conglomerates of Bârnadu. Hence, the deformation corresponds to the Austrian tectogenesis at the end of the Albian (Grasu et al. 2010, 2012).

The central-Carpathian fault unfolded in the mid-Cretaceous and it resumed in the subsequent tectogeneses, including in the Laramic phase. In this case, the Bukovinian, sub-Bukovinian, infra-Bukovinian and Transylvanian Nappes, together with the post-Austrian deposits up to the Paleocene deposits, advanced at the same time eastwards, over the flysch area (Grasu et al. 2010). In the Hăghimaş syncline, there is an allochthonous Mesozoic series, as a covering lamina for the calcareous mass within the Munticelu-Surduc Massif (Uhlig 1907; Ciocârdel and Patrulius 1960).

The Hăşmaş Transylvanian Nappe is a detachment nappe and it contains sedimentary terms accumulated on an oceanic type of crust (Săndulescu 1975). The pre-Jurassic formations of the Transylvanian sequence appear discontinuously in the wildflysch mass. The main body of the Hăşmaş Nappe has Kimmeridgian-Urgonian limestones deformed by the pre-paroxysmal tectogenetic phases (the Nevadian and Austrian-Alpine phases). The first discontinuity is placed at the Kimmeridgian-Tithonian limit, while the second at the limit between the Urgonian and the Neocomian deposits (Bojoi 2000; Săndulescu 1975; Grasu et al. 2010).

During the Austrian-Alpine phase, the folding of the calcareous mass belonging to the Hăghimaş Nappe occurred. In this case, a system of small nappes emerged (Bicaz Gorges, Hăghimaşu Negru). The force of gravity influenced the movement of the Hăghimaş Nappe. There is a synchronism between the advancement process and the sedimentation of the wildflysch formation comprising the landslide bed of the calcareous mass. The overthrust process was blocked when the limestone mass reached the axial area of the Bukovinian basin. Because of the reduced slope, the friction force managed to surpass the force of gravity. This is why the catchment basin of the wildflysch has an asymmetric profile: a smooth slope westwards and an abrupt slope eastwards. The wildflysch is absent from the sub-Bukovinian field (Săndulescu 1975). The minimum overthrust width seems to be 30 km (Săndulescu 1975) or 60–70 km (Grasu et al. 2010, 2012).

The erosion contour of the Hăghimaş Nappe is highly scalloped. The post-Austrian erosion determined the formation of tectonic klippes, tectonic windows and semi-windows, leveling nappe outliers, etc. The best-known tectonic klippes are on the eastern flank of the syncline (Munticelu, Surduc) (Ciocârdel and Patrulius 1960).

The overthrust nappe outliers were created during the detachment and advancement phases of the Hăghimaş Nappe. At the top, the pre-Kimmeridgian, Triassic and Lyassic formations that accumulated in the Transylvanian ditch were mobilized. Most leveling nappe outliers are on the internal flank of the syncline (Ciofronca, Fratele). The nappe outliers range between 10 and 300 m in size; the largest belongs to the Adneth Lyassic (Săndulescu 1975; Grasu et al. 2010).

The tectonic fault inliers (windows) demonstrate the allochthonous character of the mass of Malm-Urgonian limestones (Gropile, Cheile Mari). The tectonic semi-windows are placed on the western side and on the external margin of the main nappe body (Curmătura, Pârâul lui Lazăr, the Bicaz Gorges cabana).

The Turonian-Senonian sandstones with *Itruvia* (Beneş), in Gosau facies repose discordantly on the crystalline and on the Triassic deposits (Atanasiu 1928) and they are specific to post-tectonic deposits. The same category can include the Bârnadu Conglomerates (Vraconian-Cenomanian) disposed discordantly on the Bukovinian Nappe, on the Hăghimaş Transylvanian Nappe (limestones) and on the contact between the two Middle-Cretaceous overthrust units (Grasu et al. 2010). The post-paroxysmal deformations belong to the Laramian or Savianc, or even to the Styrian tectogenesis. They correspond to the main folding phases that affected the flysch units (Săndulescu 1975).

Most post-paroxysmal structures were not completed in one moment. They represent the inheritance of a previous deformation phase. The final moment was characterized by the retro-overflowing imposed by the compressions that intervened in the late tectogeneses. The transverse faults affect the Bukovinian Nappe and the Hăghimaş Transylvanian Nappe (Oaia brook, Cheile Mari) (Grasu et al. 2010).

It appears that the Bukovinian Nappe within the Hăghimaş syncline includes the Transylvanian nappe and other cover nappes, independently carried in its mid-Cretaceous overthrust: The Red Lake-Licaş Nappe, the Wildflysch Nappe, the Lunca Nappe, the Salamaş Nappe (Mureşan 2006, 2008). The Red Lake-Licaş Nappe includes a part of the Mezosoic sedimentary pile of the Bukovinian Nappe (sandstones and Seisian conglomerates; jaspers and Callovian-Oxfordian radiolarites; the Permian Hăghimaş breccia; red ammonite limestones of the Kimmeridgian *acanthicum*) (Mureşan 2006, 2008; Turculeţ and Grasu 1969).

The Wildflysch Nappe is a Transylvanian overthrust unit (Balintoni 1996), advanced over the Red Lake-Licaş or Salmaş Nappes (Mureşan 2006, 2008). The wildflysch was formed in the basin of the Transylvanian Tethyan, before its closing, the reason for which blocks of basic cinerites associated with radiolarites were included. During the abduction and sliding of the Transylvanian Nappes, the ophiolites au and allochthonous series were generated.

The Lunca Nappe includes the Tithonian-Neocomian Lunca Formation. It has the character of a flysch with intercalations of the Formation with *Aptychus* (synchronous from a stratigraphic perspective) but of a pelagic facies. It extends from the Dămuc basin to the Trotuş sources (Mureşan 2006, 2008). The Lunca anticline corresponds to tectonic windows, where the erosion revealed elements of the Rarău or Red Lake-Licaş sandstone Nappe. The structuring occurred during the Styrian tectogenesis (Miocene).

The Salamaş Nappe comprises the Hauterivian gravely calcareous and calcarenitic flysch (Mureşan 2006, 2008). It is a lithostratigraphic entity, more than 300 m thick, thus constituting an independent overthrust nappe, disposed on top of the Lunca Nappe, which overlays the Wildflisch Nappe.

4.3 Soil Types

The pedologic system is crucial for the overall picture, in order to understand the dynamics, causes and mutual effects among the various elements specific to the Red Lake (hydrological, vegetal, faunistic and even micro-climatic).

The soil map for the Red Lake hydrographic basin was elaborated depending on the reference layer on plate 12 (Gheorgheni) within the Map of Romanian Soils (1:200,000 scale). The reference system for this plate ("Gheorgheni"), as well as for all the other plates in Romania, is that of the SRCS (Romanian System of Soil Classification), of 1980. The paedology-related information was correlated with that of 2003, according to the new system (SRTS—Romanian System of Soil Taxonomy). It comprises a series of new elements (names, inclusion, classes of soil, etc.), compared to the abovementioned previous system (Table 4.1).

There is a close correspondence between the geologic substrate (mostly through the representative type of rock) and its soils. This connection is present in the entire basin. The great importance of the parental material and its origin can be demonstrated (Fig. 4.10).

The surface occupied by rendzinas is a lot bigger than that of limestones in the Red Lake hydrographic basin. This proves that, by desegregation, the separated

Table 4.1 Pedologic correspondences between SRCS and SRTS

Nr.	SRCS (1980)	SRTS (2003)
1	Acid brown	Dystric cambisol (cambisol class)
2	Eu-mesobasic brown	Eutric cambisol (cambisol class)
3	Rendzinas	Rendzinas (chernozem class)
4	Ferric-illuvial brown	Prepodzol (spodic soils class)
5	Brown podzols	Brown podzols (spodic soils class)
6	Rendzina eu-mesobasic brown	Sub-rendzina eutric cambisol (cambisol class)

Fig. 4.10 Rendzinas within the southeastern sector of the Red Lake

limestone material (mostly off the walls with near-vertical slopes) was rolled up onto a bigger surface, thus contributing to the extension of this soil type. The rendzinas are not very thick (20–50 cm) and given that they were formed on limestones, they developed on big slopes and these slopes do not favor a thick layer of soil.

The reduced thickness of the rendzina layer is due to the red limestone substrate. There is another argument to support the origin of the name given to the Red Lake, as, when the waters are calm, the reddish color of the limestone substrate is visible in its reflection. In the lake area—mainly on its southern margin (the outlet of the Oaia brook), where there is advanced clogging—limnic soils (specific to lacustrine basins) can be found.

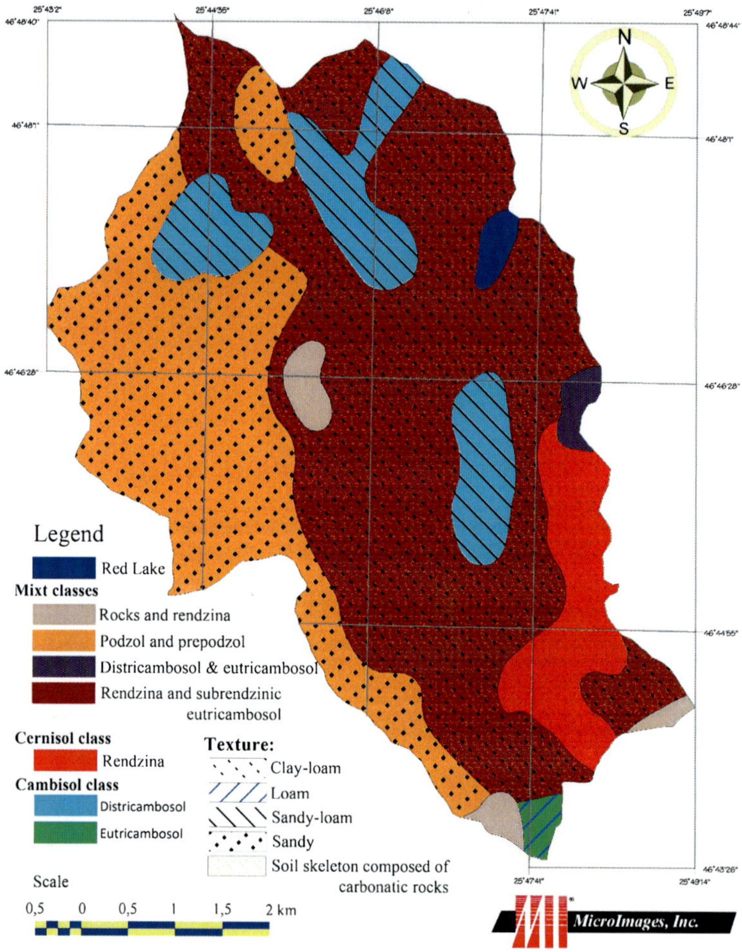

Fig. 4.11 Map of the soils within the Red Lake hydrographic basin

The rendzinas and eutric cambisols have a clayish texture. This helps us understand one of the reasons for the constant and persistent clogging of the Red Lake, given the fine size of soil particles (0.002–0.02 mm). Among the evidence, it should be mentioned the bathymetric profiles, which are mostly at the level of the tributary outlets, evolve rhythmically—the moist perimeter raises towards the surface by almost 2–4 m every year.

The podzols within the western sector developed on crystalline schists—rocks with high values of acidity (Fig. 4.11). This fact also explains the abundance of conifers, which develop on podzols. Prepodzols and brown podzols are not very fertile and they are only suitable for pastures or forested areas.

The physical support of the subjacent rocks comprises a mineral cover—through various physical–chemical processes, it transformed into the current soil cover. As it is a relatively high mountain area, the soils are not very thick or very old.

References

Atanasiu I (1928) Cercetări geologice în împrejurimile Tulgheşului (Judeţul Neamţ). Annals of the Romanian Geologic Institute 13, Bucharest

Atanasiu I (1961) Cutremurele de pământ din România. Romanian Academy Publishing House, Bucharest

Balintoni I (1996) Geotectonica terenurilor metamorfice din România. Babeş-Bolyai University, Cluj-Napoca

Băncilă I (1941) Etude géologique dans les Monts Hăghimaş-Ciuc. Yearbook of the Romanian Geologic Institute 21

Băncilă I (1958) Geologia Carpaţilor Orientali. Scientific Publishing House, Bucharest

Bojoi I (1968a) Observaţii asupra reliefului carstic din masivul Hăghimaş. In: Proceedings of the station for the biologic, geologic and geographic research Stejaru, 1

Bojoi I (1968b) Contributii la sedimentologia Lacului Rosu. In: Proceedings of the station for the biologic, geologic and geographic research Stejaru (in Romanian), vol 1. pp 87–105

Bojoi I (2000) România. Geografie fizică. Alexandru Ioan Cuza University Publishing House, Iaşi

Cărăuşu S, Ghenciu V, Munteanu A, Timofte L (1971a) Aspecte climatice hidrofizice, hidrochimice şi biologice privind Lacul Roşu din perioada septembrie 1969—august 1970. Stud Commun Museum Nat Sci Bacău 1:411–426

Cărăuşu S, Ghenciu V, Timofte LI (1971b) Unele date cu privire la trăsăturile hidrofizice ale Lacului Roşu din perioada august 1968—septembrie 1969. Sci Ann Alexandru Loan Cuza Univ Iasi 17(2):425–445

Ciaglic V (2005) Contributii la cunoasterea caracteristicilor hidrologice si hidrogeologice ale muntilor Ghilcos-Haghimas-Mezinul si a conditiilor care le determina. Terra Nostra Publishing House, Iasi, pp 1–71 (in Romanian)

Ciocârdel R, Patrulius D (1961) Observaţii asupra unei klippe de calcare jurasice încorporate depozitelor eocretacice din sinclinalul Hăghimaş. Communications of the Popular Republic of Romania Academy, Bucharest, vol 10. No 1

Dobrescu C, Ghenciu V (1970) Aspecte din vegetatia Lacului Rosu, Studii si comunicari. Museum Nat Sci Bacau 3:129–136

Dragastan O (1975) Upper Jurasic and lower cretaceous microfacies from the Bicaz valley basin. Institute of Geology and Geophysics. Memoires, 21

Gâştescu P (1971) Lacurile din Romania—Limnologie regionala. Romanian Academy Publishing House, Bucharest, pp 1–372 (in Romanian)
Ghenciu IV (1968a) Regimul de oxigen al Lacului Rosu. Communications of the Scientific Session, "Al.I.Cuza University," Iasi, pp 184–196
Ghenciu IV (1968b) Regimul termic al apei Lacului Rosu. Communications of the Scientific Session, "Al.I.Cuza University," Iasi, pp 145–160
Ghenciu IV, Apavaloaie MM (1969) Contributii la cunoasterea regimului de precipitatii din zona Lacului Rosu. Sci Ann Al.I. Cuza Univ Iasi 15(1):29–40
Ghenciu IV, Carausu S (1970) Cateva date cu privire la dinamica cantitativa a fitoplanctonului din apa Lacului Rosu in perioada iulie 1967—decembrie 1968. Sci Ann Alexandru Ioan Cuza Univ Iasi 16(1):147–163
Grasu C, Turculeţ I (1965) Asupra vârstei recifului de la Gura Dămucului. SSNG, Geology Communications, III, Bucharest
Grasu C (1971) Recherches géologiques dans le sédimentaire mésozoique du basin supérieur de Bicaz (Carpates Orientales). In: Proceedings of the Stejaru Station, Pângăraţi
Grasu C (1976) Sedimentarul mezozoic de la sud de Izvorul Mureşului. Problema Pânzei Subbucovinice. Annals of the Museum of Natural Sciences, Piatra Neamţ, Geology-Geography, vol 3
Grasu C, Turculeţ I (1980) Rezervaţia Lacu Roşu-Cheile Bicazului. Particularităţi geologice şi geomorfologice. Ocrotirea Naturii, Romanian Academy, vol 24. no 2
Grasu C, Turculet I (1998) Rezervatia Lacul Rosu—Cheile Bicazului. Particularitati geologice si geomorfologice. Ocrotirea naturii si a mediului inconjurator, 24(2):135–145
Grasu C, Miclăuş C, Brânzilă M, Baciu DS (2010) Munţii Hăşmaşului. Monografie geologic şi fizico-geografică. "Al.I.Cuza University" Publishing House, Iaşi
Grasu C, Brânzilă M, Miclăuş C, Baciu DS (2011) Parcul Naţional Cheile Bicazului-Hăşmaş. Ghidul siturilor geologice şi fizico-geografice. "Al.I. Cuza" University Publishing House, Iaşi
Grasu C, Miclăuş C, Brânzilă M, Baciu DS (2012) Sinclinalul Mezozoic Tulgheş-Hăşmaş-Ciuc. "Al.I. Cuza" University Publishing House, Iaşi
Herbich H (1878) Das Szeklerland. Budapest
Mihăilescu V (1940) Cum s-a format Lacul Roşu de la intrarea în Cheile Bicazului. Bulletin of the Romanian Geography Society, Bucharest, LIX
Mureşan M (2006) Redefinition de la zone cristallino-mesozoique des Carpates Orientales et son charriage post-autrichien sur la zone du flysch. Annals of the Romanian Geologic Institute, vol 74. Bucharest
Mureşan M (2008) Nappes de charriage de couverture alpines dans la zone cristallino-mésozoïque de Carpathes Orientales. Un exemple: la région des Monts Hăghimaş. Annals of the Romanian Geologic Institute, vol 75. Special Issue, Bucharest
Mutihac V (1990) Structura geologică a teritoriului României. Technical Publishing House, Bucharest
Pandi G, Magyari ZS (2003) Realizarea hartilor batimetrice pe calculator. Modelul Lacul Rosu. Studia Universitatis Babes-Bolyai, Cluj, pp 55–60
Pandi G (2004) A Gyilkos-Tó. Hidrogeográfiai tanulmány. Casa Cartii de Stiinta Publishing House, Cluj, pp 1–231
Pandi G, Buzila L (2004) Caracteristici hidro-geomorfologice ale sedimentarii in Lacul Rosu. Geography within the context of contemporary development, Cluj-Napoca University Press, Cluj
Pelin M (1967) Studiul geologic al formatiunilor mezozoice din Masivul Haghimas (intre Lacul Rosu si Piatra Unica). Ph D thesis, Bucharest
Pelin M (1971) Consideratii asupra tectonicii regiunii Lacul Rosu—Piatra Unica. Annals of the Bucharest University, vol 20. Bucharest, pp 64–76
Pisota I, Nastase A (1957) Lacul Rosu, nod de confluenta a trei bazine hidrografice. Probleme de geografie (in Romanian) 4:181–205
Popescu GA, Dimitriu A (1950) Observatiuni piscicole la Lacul Rosu. Bulletin of the Institute for Piscicle Researches, vol 9. Bucharest, pp 33–47

Popp N (1941) Cronica Geografica, A) Fapte. Congresul Profesorilor de Geografie de la Piatra Neamt. Bulletin of the Romanian Society of Geography, Bucharest, LX, pp 317–342 (18–21 May 1940)

Preda I, Pelin M (1963) Contribuţii la cunoaşterea geologică a împrejurimilor Lacului Roşu (Carpaţii Orientali). Society of Natural and Geographic Sciences in Romania, Geology Communications, Bucharest 2:107–116

Preda I (1967) Deplasari de teren in zona Lacului Rosu. Geology Communications, vol 4. Bucharest

Preda I (1969) Consideraţii asupra tectonicii Masivului Hăghimaş. Bulletin of the Geology Society, vol 9. Bucharest

Preda I (1971) Consideratii hidrogeologice asupra Muntilor Haghimas. Bulletin of the Romanian Society of Geologic Sciences, vol 13. Bucharest, pp 92–106

Preda I (1976) Contribuţii la cunoaşterea Liasicului şi Doggerului din Munţii Hăghimaş (Carpaţii Orientali). Yearbook of the Museum for Natural Sciences Piatra Neamţ, Geology-Geography, vol 3. pp 19–41

Puscariu V (1939) Lacul Rosu si Cheile Bicazului. Touring Clubul Romaniei Publishing House, Cluj

Romanescu G (2009a) Trophicity of lacustrine wetlands on the Carpathian territory of Romania. A case study from the East Carpathian mountains. In: Proceedings of the "Dimitrie Cantemir" Geographic Seminar, vol 29. pp 5–13

Romanescu G (2009b) Trophicity of lacustrine waters (lacustrine wetlands) on the territory of Romania. Lakes Reservoirs Ponds 3:62–72

Romanescu G (2009c) The physical and chemical characteristics of the lake wetlands in the central group of the east Carpathian Mountains. Lakes Reservoirs Ponds 4:94–108

Săndulescu M (1975) Studiul geologic al părţii central şi nordice a sinclinalului Hăghimaş (Carpaţii Orientali). Yearbook of the Romanian Geologic Institute, Bucharest, XLV

Senchea N (1948) Formarea lacurilor din Romania. Vasile Adamachi Sci J Iasi 24: 46–52

Turculeţ I, Grasu C (1969) Asupra existenţei Stratelor cu Aptychus în regiunea Tulgheşului. In: Proceedings of the "Stejarul" Station, Pângăraţi, 2

Udriste O (1963) Lacul Rosu si imprejurimile. Meridiane Publishing House, Bucharest, pp 1–55 (in Romanian)

Uhlig V (1907) Uber die Tektonic der Karpathen. Sitzungsber. k. Akad. Wiss, XVI, Vienna

Xantus I, Xantus J (1999) Hagymas-hegyseg es a Gyilkos-to kornyeke. Pallas Akademia, Csikszereda, pp 1–168

Zaruba Q, Mencl V (1955) Geologie inginerească. Prague

Chapter 5
Morphographic and Morphometric Features of the Hăghimaş Mountains, the Hydrographic Basin and the Red Lake Lacustrine Basin

Abstract The tectonics of the erosive substrate (composed of limestones and conglomerates) is highly fragmented and represents an important factor in the morphology and morphometry of the landscape. The hydrographic basin is individualized by a series of variables, from which the most important are: rocks; soil layer and vegetal layer. The morphometric parameters for the Red Lake hydrographic basin were determined through the computer-based cartography programs TNTMips (Microimage) and ArcGis (ESRI). Out of these, the most important are: basin length, width, drainage network length and frequencies, basin surface, landform orientation, slope exposure etc. The Horton-Strahler hierarchy of the hydrographic network was also assembled, in order to point out the maximum order number of the basin river system and make a statistical analysis on the Red Lake tributaries of each Horton-Strahler river order. A detailed analysis (from both a spatial and a statistical point of view) was made on different hydrographic basins, according to the corresponding river's Horton-Strahler order. The lake basin parameters were measured with state of the art equipment (LEICA TCR 1201 total station, together with LEICA GPS 1200; Valeport Midas Surveyor echo sounder), which was used to obtain a bathymetric map of the lake and the morphometry of the surrounding areas. The bathymetric map was used to generate transects, which would later be used in silting analysis and other detailed analyses.

5.1 The Hăghimaş Mountains

The fragmentation degree of the Hăghimaş Mountains is relatively high because of the tectonics of the erosive substrate (limestones and conglomerates). This is why they are easy to cross on the east–west alignment (Moldavia-Transylvania). The passes do not exceed 1,256 m in height: Borsec (Creanga) 1,111 m; Tulgheş (Ditrău) 1,025 m; Pângăraţi (Bicaz) 1,256 m; Ghimeş (Pucioasa) 1,159 m.

G. Romanescu et al., *Limnology of the Red Lake, Romania*,
DOI: 10.1007/978-94-007-6757-7_5,

The morphological heterogeneity of the landform is determined by a sequence of rocks and by a high degree of erosion tendency. An extremely important factor is that of time. The landform characteristic is given by the dominant presence of Triassic dolomites and limestones, which make a relatively imposing landform, with isolated peaks, canyons, etc. The rounded crests, with altitudes up to 1,500 m, are formed on crystalline rocks.

High plateaus with quasi-horizontal or low-inclined slopes give a specific note to the landform in the Hăghimaş Mountains. These units are underlined by the existence of saddles. Higher altitudes are maintained along the dolomitic mass within the inside flanks: Licaş 1,675 m; Calului 1,625 m; Ciofronca 1,607 m; Suhardul Mare 1,507 m. The high altitudes in the east are due to calcareous shreds: Suhardul Mic 1,310 m; Bardosul 1,289 m; Ghilcoş (Ucigaşul) 1,406 m; Hăghimaşul Negru 1,773 m; Hăghimaşul Mare 1,792 m; Piatra Singuratică 1,587 m; Fratele 1,671 m, etc. (Fig. 5.1).

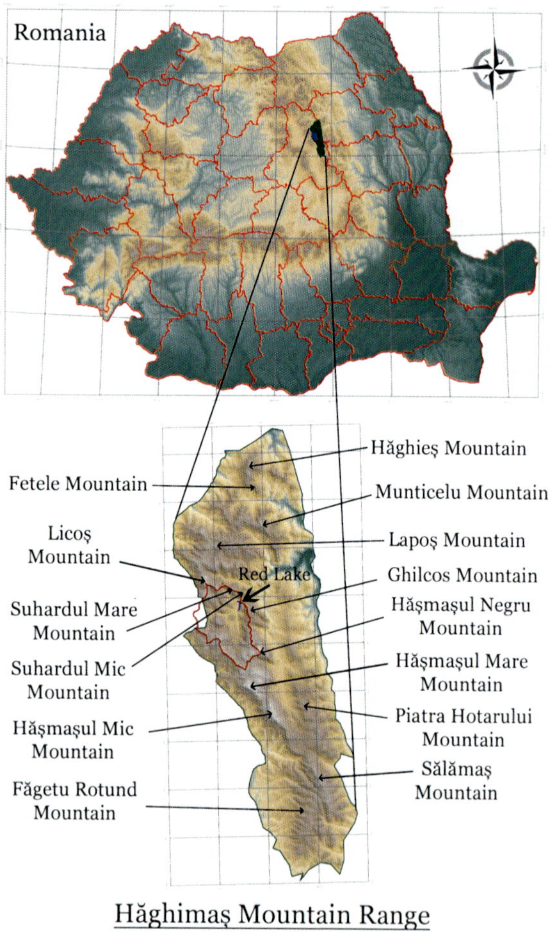

Fig. 5.1 Geographic location of the Hăghimaş mountain range and its main massifs

Among the depressions, the following are underlined: Oaia (Oii) brook-Red Lake; Trei Fântânele-Bicăjel; Bălan-Sândominic (gulf of the Cicu depression). The lower altitudes (900–1,000 m) and the elongated form isolate them from the mountain units surrounding them. These are depressions of lithological and structural contact (Oaia-Roşu and Fântânele-Bicăjel), grafted on loamy-ruditic clays of the Cretaceous or tectonic contact wildflysch (Bălan-Sândominic).

The *petrographic* landform is given by the nature of the modeling rocks and they create the following petrographic associations: crystalline schists; Dogger petrographic association; calcareous-dolomitic petrographic ensemble; Cretaceous wildflysch complex; post-tectonic conglomerates (Bojoi 1976).

From this point of view, for the Red Lake, the Dogger petrographic association and the Cretaceous wildflysch complex are underlined. The Dogger petrographic association mainly develops in the Red Lake area. Middle Jurassic deposits have a great variety of rocks and it is 200 m thick (micrites; quartz siltitic microsparites; biomicrites and quartz siltite spongolitic biomicrosparites; spongolitic loams; spongolitic limestones; calcareous spongolites; loamy sparites; and calcareous litharenites) (Grasu et al. 2010).

The landform has shapes with a conic profile (Suhardul Mare), strongly inclined slopes (20–30°), valleys with a sharp "V" transverse profile and a longitudinal profile with numerous slope ruptures (Suhard brook). The Cretaceous Wildflysch has a rather diversified petrographic constitution: clays with blocks and flysch.

The typical lithofacies contains a non-stratified clay mass, with intercalations of tilloid conglomerates, polymictic conglomerates, breccias and micro-breccias. The clay mass also includes exotic klippes (olistoliths) made from crystalline schists, Triassic dolomite, Tithonian-Neocomian and Urgonian limestones, amygdaloidal basalts, red clays, sandstones, jaspers and radiolarites, manganiferous siderite limestones, basic cinerite, etc. (Grasu et al. 2010).

The flysch paratypic lithofacies is made of loams and stratified clays, fissured limestone sandstones, muscovite sandstones, breccias and stratified calcareous or polymictic conglomerates (Grasu et al. 2010). It also comprises sedimentary klippes with limestones (50 %) of 15–20 m^3 blocks and small limestone massifs of over 10,000 m^3 (Săndulescu 1975).

The floor of the complex is the place where the Trei Fântâni-Bicăjel and Bardos-Bârnad depressions evolved, as well as the deep saddles of Suhărzel, the Red Lake (Suhardul Mare), Stâna, etc. For the crystalline schists, the epi-metamorphites of the Tulgheş group and the mesometamorphites of the Bretila-Rarău Group are specific. From a petrographic perspective, they contain phyllites, amphibolites, mica schists, gneisses, diorites, gneiss granodiorites and chloritoid, sericite and quartzitic schists (Bojoi 2000; Catana 1989).

The landform developed on the crystalline schists ranges between 800 and 1,400 m, on the western and eastern flanks. The peaks (Mediaş-Pângăraţi, Dămuc) are shaped as widely rounded shields (Bojoi 1976, 2000; Ciaglic 2005). The valleys are deep, with highly inclined slopes and a sharp "V"-shape profile.

The area of the Hăghimaş Mountains is characterized by the existence of the calcareous-dolomitic complex of 100 km^2. Unfortunately, this complex is compartmented into three subzones: central, with a longitudinal orientation, with high altitudes; eastern, corresponding to the Dămuc-Criminiş-Chicera peak, with island-shaped dolomites; western, with elongated dolomitic massifs (Licaş, Danturaş, Calul, Ciofronca) (Bojoi 1976; Grasu et al. 2010).

The Bârnadu post-tectonic conglomerates outcrop on large surfaces north from the Bicaz Gorges and in the Bicăjel basin. They are Vraconian-Cenomanian and they modify elements within the Bukovinian series and the Hăghimaş Nappe (crystalline, Middle-Jurassic limestones, loams, limestones within the Formation with *Aptychus*, Kimmeridgian-Urgonian limestones, Triassic dolomites, jaspers, diabases, vacuolar basalts, etc.). The Jurassic-Neocomian and Urgonian klippe limestone blocks register volumes between 200 and 50,000 m^3 (Săndulescu 1975).

The geomorphologic processes are dominated by chemical erosion, desegregation, debris overflows and slides (Rădoane and Rădoane 1986). The phenomena are also favored by the significant amount of precipitation.

The conglomerate-specific landforms are usually longitudinal peaks affected by selective erosion, with slopes ranging between 20 and 30°, detached from the friable deposits. The peaks and hummocks are conical (Ciurgău 1,412 m, Lapoş 1,448 m, Toşorog 1,436 m, Hăghieş 1,505 m). The valleys have a sharp "V"-shape profile (Bicaz Valley downstream from Cheile Mari).

As regards the *structure*, the area ended its evolution during the Austrian tectogenesis (Albian). There were intense ruptures and local thrusts in the rigid compartment of the Hăghimaş Nappe during the post-mid-Cretaceous tectonic movements, leading to the creation of a flake-like structure.

The landforms imposed by the structure are the following: the landform of the Hăghimaş Nappe; the structural plateaus; the tectonic window and semi-window landforms; structural steep high cliffs; and klippe landform (Bojoi 1976).

The front of the nappe is specific to the massifs of Surduc, Ghilcoş (Ucigaşu), Hăghimaşu Negru, Telecu Mare, Hăghimaşu Mare and Fratele. It is 20 km long and it has differences in altitude ranging between 400 and 600 m. Between the Trei Fântâni-Bicăjel Depression and the high peaks, the contact is lithological and structural; the contact between the Bukovinian Nappe and the Bukovinian Autochthon must be also mentioned (Ciaglic 2005).

The slopes, in levels of 50–100 m, exceed 25° and intense geomorphologic processes affect them. The erosion is intense on the direction of transverse dislocations (Pârâul lui Lazăr, Jgheabul Rotund, Pârâul Sec).

Structural plateaus are typical for the calcareous-dolomitic areas. They have fragmented surfaces and, most of the times, they are inclined or quasi-horizontal (Telec, Suhărzel). Rarely, the angles exceed 20–30° (Fratele) (Bojoi 1987, 2000).

The Hăghimaşul Mare plateau is 2 km long and 300–400 m wide. The multitude of longitudinal and transverse faults managed to compartment the plateau. There are small-sized structural plateaus in Ghilcoş (Ucigaşu) as well as Hăghimaşu Negru (Bojoi 1976; Ciaglic 2005; Grasu et al. 2010, 2012).

The landform shaped as a tectonic window or semi-window is due to the post-overthrust phase, after the accumulation of the Bârnadu Conglomerates.

The geomorphologic processes facilitated intense erosion and the resurfacing of the autochthon (the Wildflysch Formation).

The tectonic windows (Gropile, Cheile Mari) and semi-windows (the area of the Bicaz Gorges cabin, Suhardul Mic, Suhardul Mare, Surduc, Ghilcoş, Oaia Brook, Stâna Curmătura, Poiana Albă) are represented by depressions and gulf-depressions (Pârâul lui Lazăr, Pârâul Sec, Jgheabul Rotund), deep saddles and calcareous shreds (Bojoi 1976, 1988; Ciaglic 2005; Grasu et al. 2010, 2012).

The structural steep high cliffs may vary in size by nature of the rocks and by magnitude of the tectonic movements or of erosion processes. The western steep high cliff of the Hăghimaş Mountains has the appearance of a *cuesta*. The twofold *cuesta* can also appear as a hogh-back as it cuts the limestone and dolomite deposits with a 45° angle (Bojoi 1976).

The landform made of tectonic klippes includes the isolated calcareous massifs reposing over the wildflysch siliciclastic formation (Bojoi 1976, 2000; Ciaglic 2005; Grasu et al. 2010, 2012). They have a tectonic origin because they separated from the Hăghimaş Nappe during the overthrust or slides of blocks synchronous with the accumulation of wildflysch (sedimentary klippes) (Săndulescu 1975). They can be preserved in the landform also because of differential erosion (Munticelu, Piatra Luciului, Piatra Glodului) (Grasu et al. 2010).

The *karstic* landform is specific to areas dominated by limestones and dolomites. Here both types of landform are present: exokarstic (karren; sinkholes; uvalas; poljes; saddles; karstic depressions) and endokarstic (underground riverbeds; swallow holes; suspended tunnels; caves).

The most spectacular and attractive landform within the Hăghimaş Mountains is represented by the karstic elements. This is why it has been intensely studied by geomorphologists, geologists and hydrogeologists: Bojoi (1968a, b, 1970a, b, 1971a, b, c, 1972), Ciaglic (2005), Grasu et al. (2010, 2012), Mayer (1932), Naum and Preda (1972a, b), Popp (1941), Preda and Naum (1973), etc.

The most important deposits that could turn into karsts are made of the limestones belonging to the Transylvanian (Kimmeridgian-Urgonian) Autochthon and the dolomites belonging to the Bukovinian (Triassic) Autochthon. The Triassic comprises Campilian limestones with layers up to 30–35 m thick. From the perspective of the microfacies, they comprise micrites, sparites and quartz-siltite-sericite microsparites. The insoluble residue has values ranging between 2 and 25 %, with the rest comprising carbonates.

The calcareous-dolomitic layered dolomites are 40–50 m thick, with a significantly reduced insoluble residue (1.5–7 %). The massive Anisian dolomites are 150–200 m thick. From the perspective of the microfacies, they are dolo-micrites; chemically, they are dolomites and calcareous dolomites. They present an insoluble residue of 0.8–3.6 % (pure dolomites) (Grasu et al. 2010).

The Hăghimaş Nappe, 40 m thick, comprises a unit of red nodular limestones, a unit of grey-greenish limestones and another of pelitoidal sandstones. From the perspective of the microfacies, it comprises biomicrites, biopelsparites, quartzite-siltitic biopelmicrites, slightly clayish gravel and quartzite-siltitic limestones. There is 15 % pelitic material (Grasu et al. 2010).

The Stramberg Tithonian limestones are 200–300 m thick and they have a large petrographic variety. From the perspective of the microfacies, they are made of biomicrites, biopelsparites, bioopelsparites, intrasparites, intramicrites, etc. The values of the carbonates range between 91 and 99 %, which means they are practically pure (Grasu et al. 2010).

Neocomian has two facies: pelagic, with pelitomorphic limestones and loams with ammonites and tintinnids; and neritic-reef with *Leviathania*. From the perspective of the microfacies, it contains micrites, quartz-siltitic biomicrites and clayish biopelmicrites. The content of carbonates ranges between 60 and 95 % (Grasu et al. 1995).

The layers of Urgonian and Barremian-Aptian limestones are up to 200 m thick. They have a reef origin and they comprise biomicrites, biopelsparites and bioopelsparites. They contain epiclastic material ranging between 0.28 and 1.85 %.

The karstification speed was calculated based on Corbel's formula (1957) (Table 5.1). In ths sense, other areas in the world have higher values than those in the Hăghimaş Mountain: France 129 m^3/km^2/year, Slovenia 109 m^3/km^2/year, Jura Mountains 90 m^3/km^2/year, Bosnia 90 m^3/km^2/year (Bojoi 1970a, b, 1971a, b, c). With close values to the Hăghimaş Mountains, Dalmatia, with 28 m^3/km^2/year and Poland 20 m^3/km^2/year should be mentioned (Bojoi 1970a, b; Ciaglic 2005).

The karstification values include the Hăghimaş Mountains in the category of average (between 20 and 50 m^3/km^2/year) towards slow activity (under 20 m^3/km^2/year).

Table 5.1 Karstification speed in the area of the Hăghimaş Mountains (after Bojoi 1970b, 1971a, c; Grasu et al. 2010)

Source	Altitude (m)	Emergence rock	Carbonate content (%)	Discharge (l/s)	Volume of dissolved limestone (m^3/km^2/year)	Ablation speed (cm/1,000 years)
Bicaz river	718	Limestone	–	4.28	23	2.3
Paleu source	1,091	Sandstone	40.75	2.00	30	3.0
Surduc source	720	Limestone	98.25	6.00	24	2.4
Şugău source	714	Limestone	94.00	3.50	27	2.7
Tunzărie source	1,219	Wildflysch	78.50	1.30	22	2.2
Melekviz source	1,012	Limestone	98.50	162–309	21	2.1
Trei Fântâni source	1,074	Limestone	98.50	130	22	2.2
Piatra Calului source	1,057	Dolomite	86.25	3.12	19	1.9
Poiana Luminişului source	1,146	Dolomite	92.00	28	20	2.0

The exokarstic relief presents different forms by the nature of rocks (limestone or dolomite) and by the positions that they occupy in the territory. Because of the different ages, degradation and attenuation, the exokarstic relief is not representative (Bojoi 1968a, b).

Most lapiés are present on structural surfaces and on low-gradient slopes. Most lapiés are rain-generated lapiés, mixed lapiés, tabular lapiés (clints) and linear lapiés (Bojoi 1971a, b, c; Săndulescu 1975). The most common are the mixed lapiés, forming fields with an irregular micro-relief. In this case, the tabular and linear lapiés are associated to the Bardos brook and to the Ghilcoş Mountain.

Tabular lapiés are deep, with a circular or oval section, with a diameter up to 10–15 cm (Colţul Mare Mountain, Surduc Mountain). Linear lapiés vary in depth, from a couple of centimeters up to 1 m and in length from 2 to 10 cm. The flutes on dolomitic slopes intercalate in acute angles, which leads to the creation of a network with rhombic holes. The rain-generated lapiés are due to the impact of raindrops with the friable substrate. The small resulting excavations have an irregular contour and sharp margins, thus acquiring a honeycomb-shape (Cheile Mari, Cheile Şugăului, Cheile Cupaşului). Most lapiés are masked by more or less thick diluvial masses or soils in various stages of evolution (Bojoi 1968a, b; Ciaglic 2005; Grasu et al. 2010).

Dolines appear on structural plateaus and on gently inclined slops, covered by calcareous and dolomitic deposits. They can develop on both an uncovered karst-favorable substrate and on a substrate covered by diluvial or clastokarst layers.

From a genetic perspective, the studied area comprises all types of dolines within the Central-European karst: corrosion dolines, erosion dolines, crevice dolines, corrosion and suffusion dolines, corrosion and erosion dolines, suffusion and collapse dolines, polygenetic dolines. Size wise, they are part of the following categories: very small (diameter and depth down to 3 m—Javardi basin, Telecul Mic, Curmătura, Ciofronca), small (3–s 10 m—Sacadat-Piatra Crăpată, Hăghimaşul Negru, Bardos basin), medium-sized (10–15 m—Fratele, Curmătura), large (200–400 m—Poiana Albă, Pârâul sec) (Bojoi 1971a; Grasu et al. 2010, 2012).

Most dolines are part of the following typology: dolines with circular or oval plan geometry and a funnel-shaped profile; dolines with an elongated channel and a "V" or "U" cross-section; flat-bottomed bowl and a flat and gently inclined framework; well-shaped dolines, with strongly inclined walls (Bojoi 1971a; Grasu et al. 2010, 2012).

Most small dolines are funnel-shaped, with circular plan geometry (Poiana Albă, Hăghimaşul Negru, Fratele, Sacadat-Piatra Crăpată). The funnel-shaped dolines with an oval plan geometry are solitary (Curmătura saddle, Cupaşul basin) (Bojoi 1971a; Ciaglic 2005). They have regular shapes and a slow evolution.

Dolines with an elongated channel and a "V" or "U" cross-section accompany fissures reaching the surface of karst-favorable deposits on strongly inclined walls. The corrosion-erosion processes widen them. In the absence of collapse

Table 5.2 Density of the dolines in the Hăghimaş mountains (after Bojoi 1971a, b, c)

Karstic zone	Surface (km^2)	Density (number of dolines/km^2)
Sacadat-Piatra Crăpată	3.1	3.5
Fratele mountain	9.1	3.5
Stâna Curmătura	1.9	1.0
Hăşmaşul Mare	13.3	0.9
Hăşmaşul Negru	13.7	1.1
Ghilcoş (Ucigaşu) mountain	8.3	0.7
Surduc-Suhard-Bardos	8.1	1.2
Licaş mountain	5.9	1.0
Javardi basin	3.3	1.2

phenomena, the cross-section acquires a "V" profile. When collapses occur, the cross-section profile becomes a "U" (Licaş, Fratele, Colţul Mare). The flat-bottomed dolines characterize the large landforms. They are formed because of dissolution of underground transportation. They usually have a swampy bottom, caused by the clay specific to the decalcification of limestones (Hăghimaşul Negru, Ghilcoş, Poiana Albă) (Bojoi 1971a). Well-shaped dolines, with strongly inclined walls on the vertical, usually continue in depth through a swallow hole (Licaş).

The density of the dolines within the Hăghimaş Mountains is relatively low: <1 doline/km^2 in the dolomite-calcareous area; 3.5 dolines/km^2 in the sectors of the Malm-Urgonian limestones (Table 5.2). Surface wise, they range between 1.5 and 13.7 km^2. Their linear placing demonstrates the existence of subterranean streams.

The uvalas within the Hăghimaş Mountains are penetrated and drained by surface brooks. In the low sector, they may present swampy zones or peat accumulations. They can reach 400 m in length and 100 m in width. The best-known uvalas can be found in the following areas: Sacadat-Piatra Crăpată, Poiana Albă, Hăghimaşu Negru (Bojoi 1971a, b, c).

The poljes within the Hăghimaş Mountains fall into two categories: plateau poljes (Sacadat-Piatra Crăpată, Stâna Curmătura, Poiana Albă, Ciofronca-Pârâul sec) and marginal poljes (Pârâul lui Lazăr, Pârâul Sec, Pârâul Jgheabul Rotund). Most poljes reach 5 km in length and 1 km in width (Poiana Albă) (Bojoi 1971a, b, c).

The endokarstic relief comprises a multitude of subterranean holes caused by corrosion, mechanic erosion and collapses: subterranean riverbeds, swallow holes, terraced tunnels, caves. Through the endokarstic relief, the subterranean supply contributes to maintaining a hydrologic balance of the rivers and lakes in the area (Puşcarciuc 1988).

The subterranean riverbeds appeared because of the erosion and corrosion of the fissures within the calcareous mass; they have high-discharge emergence. Most of the times, the discharges are intermittent, related to the amount of precipitation fallen in that area. Up to the present, the following subterranean riverbeds have been identified and mapped: Melekviz, Surduc, Bardos, Ciofronca and Trei Fântâni (Bojoi 1971b; Ciaglic 2005; Grasu et al. 2010). Most of them are permanent, though the discharges may decrease considerably. The discharges of the subterranean streams range between 10 l/s and 300.400 l/s during rainfall. In the absence of precipitation, the discharges drop significantly or the streams simply dry out (Bardos).

The most frequent endokarstic landforms are represented by swallow holes. Water circulation on the vertical created elongated landforms. From a dynamic perspective, swallow holes fall into the following categories: current, active, fossil and inactive.

The most frequent active swallow holes make the connection between the superficial and subterranean forms (Licaş, 8 m in diameter and 50 m in depth). Most inactive swallow holes are on the western steep cliffs of Hăghimaşul Mare, Piatra Roşie, Fratele, Poiana Albă (4 m in diameter and 6 m in depth), Telecul Mare (2 m in diameter and 20 m in depth), etc. (Bojoi 1970b, 1971b, Ciaglic 2005; Giurgiu 1990; Grasu et al. 2010).

Terraced tunnels appear in old subterranean riverbeds intersected by superficial riverbeds (Javari brook). They are small-sized and their dynamic depends on rock hardness. The existence of these tunnels is associated with the existence of riverbeds, therefore they can be an indicator of the presence of some of the riverbeds.

Though the karstic relief of the Hăghimaş Mountains is relatively well developed, caves are not specific to the area. Along the subterranean holes, there are seven caves, six of which are limestone and one of dolomites: two in the Bicaz Gorges, three on Pârâul Sec (tributary of Bicăjel), one in the Javardi Gorges (Fratele Mountain) and one on the left slope of Suhard (dolomites) (Bojoi 1968a, b, 1971b; Giurgiu 1990; Grasu et al. 2012; Orghidan et al. 1965). The best known are as follows: Peştera cu Izvor (20 m), Peştera cu Lublinit (35 m), Peştera cu Fereastră (35 m), Peştera Urşilor (17 m), Peştera Toşorog, Peştera Munticelu.

The term *cave* is somewhat inadequate, as they are small-sized and they are not "endowed" for proper speleological study. The only real caves are Toşorog and Munticelu. Subterranean galleries were found, with no contact to the surface.

It is known that, before the creation of the Bicaz Gorges, there were a series of karstic depressions down to 1,000 m deep. In the current appearance of the valley, one can still distinguish the contour of three karstic depressions. The depression currently filled by the Red Lake may have been an old polje that touched the karstification base in the Quaternary and that the Bicaz River drained subsequently (Bojoi 1968a, b, 1970a, b, 1971a, b, c, 1972, 1987).

The fluvial relief is developed along the main streams, with a current or old itinerary. Eighty percent of the fluvial landform belongs to the Bicaz hydrologic system and the rest to the basins of Bistricioara, Olt and Trotuş (Bojoi 1971a, b; Grasu et al. 2010).

The Bicaz Valley is segmented into three distinct sectors: western (the origin), coinciding with the stream of the Oaia (Oii) brook; central (which laps the Bicaz Gorges); eastern (from the gorges' exit to the contact with the flysch). The essential features are given by the dominance of the erosive processes and the weak development of the accumulative ones.

The energy of the longitudinal profile of Bicaz River is 865 m; 300 m correspond to the sector situated between the Red Lake and the confluence with the

Şugău brook (8 km). Within this sector, there are four slope ruptures: the Red Lake outlet, at the level of the former Bicaz Gorges cottage, downstream from Piatra Altarului (Cascada Bardosului), at the confluence with Bicăjel (Gâtul Iadului).

On the Bicaz valley, alluviations occur in the low-gradient sectors (<4 %) and the rocks are eroded directly in the high-gradient sectors (<4 %) (Bojoi, 1970a, 1971b). Most valleys within the karstic sectors present riverbed marmites or lateral suspended marmites, sometimes at considerable heights (Bicaz Gorges, Şugău Gorges, Bicăjel Gorges, Lapoş Gorges, etc.). The depression areas along hydrographic arteries present wide, meandered, or even interlacing riverbeds, accompanied by the development of the alluvia-specific landforms (along Bicaz, the depression area of the Oaia brook, the Bicăjel brook, the Lapoş brook, the Piatra Roşie brook, etc.).

The best-developed terraces are unfolded along the Bicaz River and some of its tributaries (Dămuc, Jidanului, etc.). On the Bicaz Valley, eight terrace levels have been identified: 4–8, 8–10, 15–20, 35–40, 40–45, 60–65, 75–85, 110–120 m (Bojoi 1971a, b, c). On some of its tributaries, five terrace levels have been delimited: 3–5, 5–7, 8–10, 15–20, 30–35 m (Dămuc).

The 3.5 km long Bicaz Gorges have a known complex development. The initial, epigenetic phase was followed by a karstic phase, when important hydrographic alterations occurred. The deepening phase took place in the deposits of wildflysch with Malm-Urgonian limestones (the Hăghimaş Nappe) (Bojoi 1970a, b, 1972, 1988, 2000).

The current morphologic facies is also influenced by the existence of the ancient periglacial (cryonival) processes. The landforms can be sculptural (after congelifraction; ditches for evacuating the desegregation debris; couloirs) and accumulative (stone fields; debris; wind accumulations; solifluxion deposits) (Bojoi 1976).

Because of reduced altitudes, the Hăghimaş Mountains have only gone through periglacial modeling in the cold Pleistocene climate. In this case, the term *periglacial* is used to designate the morphogenetic processes within the cold climate. The current periglacial processes in the Hăghimaş Mountains are significantly weaker than during the Pleistocene. Landform modeling during the cold period of the Pleistocene was possible through gelifraction (temperatures <0 °C). The most frost-riven rocks were the limestones and the dolomites. The desegregation of these deposits was facilitated by the existence of numerous fissures and fractures (Bojoi 1969; Grasu et al. 2010, 2012; Ichim 1973, 1976).

The gelifraction, the transportation of desegregation products, the accumulation of detritus and other detached materials, etc. represent the main characteristics of the periglacial and of the area corresponding to the Hăghimaş Mountains. The most important periglacial landforms are sculptural (congelifraction witnesses—columns, towers, pillars, detritus runways, avalanche couloirs), accumulative (rock fields, detritus, solifluxion deposits, eolian deposits) and sculptural-accumulative.

The geomorphologic division into zones of the Hăghimaş Mountains is based on the association between types, subtypes and landforms (Bojoi 1969, 2000):

- The unit of high calcareous plateaus with a structural character (the Ghilcoş [Ucigaşul] axial plateau, Hăghimaşul Negru, Munticelu, Javardi, Hăghimaşul Mare, Telecul Mare, Fratele, Suhard, Cupaş, Surduc, Bardos);
- The unit of steep high cliffs developed on limestones, dolomites and conglomerates, comprising the following subunits: Hăghimaşul Mare hogh-back; the steep slopes accompanying the valleys within the gorges (Bicaz, Bicăjel, Cupaş, Munticelu, Lapoş); the eastern steep slope towards the Depression Trei Fântâni-Bicăjel (the front of the Hăghimaş Nappe); the steep slopes of isolated, calcareous, dolomitic and conglomeratic massifs (Munticelul, Hăghieş, Danturaşul, Licaş, Vithavaş);
- The unit of insular mountains, with conical peaks, within the north–western and western sectors;
- The unit of low crests, with north–south orientation, overlaying crystalline schists on the margin of the syncline Mesozoic sedimentary (towards the depressions of Giurgeu and Ciuc);
- The unit of low crests, on the outer side of the Hăghimaş Mountains, structurally overlaying the Dămuc crest (and its southward prolongation), on the Luncat anticline, sculpted in the crystalline and on the sedimentary continuation (from the Triassic dolomites to the Cretaceous wildflysch);
- The unit of intra-mountain depressions: Trei Fântâni-Bicăjel, Oilor, Bârnad, Upper Suhard, Toşorog (on the bottom of the Jidanului valley).

5.2 The Red Lake Hydrographic Basin

A hydrographic basin is considered an open system, where there are matter and energy exchanges with the environment. Precise delimitation, with the help of cutwater, makes the quantitative studies exact, as well as correct estimations of the energetic values and matters coming in and out of the basin. The hydrographic basin is a whole, submitted to a continual flow of matter and energy. The current morphology of the hydrographic basin is the result of a long evolution process, during which a dynamic balance was achieved between the variables defining the behavior and general flow of matter and energy (Zăvoianu 1978).

The factors acting within a hydrographic basin are represented by subaerial agents that, through the flow of matter and energy introduced in the system, act continuously to lower the surface. The opposing factors provide unity for the entire basin, in a continuous adaptation process to achieve a dynamic balance.

Among the variables contributing to the individualization of the basin, the following must be underlined: rocks; soil layer; vegetal layer. They all form a well-defined totality, quantitatively speaking. For the morphological analysis of the catchment basin, geography-specific means were used: topographic maps with a 1:25,000 scale; geological and land use maps. The topographical maps were reambulated in 1982.

Fig. 5.2 The use of total station for the topographic elevation of the lacustrine basin

In order to generate the maps by theme and to execute the measurements on aerial and satellite images, the (licensed) TNTMips program was used. The data were processed in the Geoarchaeology Laboratory within the Faculty of Geography and Geology of the "Al.I. Cuza" University, Iasi. The topographic elevations were done with the help of the Total Station (Fig. 5.2).

The morphometric parameters for the Red Lake hydrographic basin were determined through the computer-based cartography programs TNTMips (Microimage) and ArcGis (ESRI). The methodology of generating the land declivity maps, as well as the maps for slope exposition, hypsometry, drainage hydrographic network and hydrographic basins of different orders, landform fragmentation depth and density of the hydrographic network implied several techniques. They are the following: *scanning of the topographic support* at a 1:25,000 scale (DTM, 1982 edition) for the area of interest; *map importation* in the computer-based cartography program TNTMips (Microimage); *georeferentiation of topographic maps* in the Universal Transverse Mercator system (35 Area—*E* 24 *at E* 30, *WGS* 1984); *theme editing* of vector layers *t* on-screen. The vector layers obtained contain contour lines with 5 m equidistance, the drainage hydrographic network and hydrographic basins of different orders. After vectorizing the specific vector layers, they were classified in order to be analyzed quantitatively and qualitatively. The contour lines were classified depending on the altitude (m). They represented the grounds for the non-classified hypsometric map. The drainage hydrographic network was classified according to Horton-Strahler hierarchy (representing the basis for the vector layers belonging to basins of different orders, ending with the value 6). The hypsometric map of the Red Lake hydrographic basin plays the most important role as it constituted the basis for raster layers by theme referring to land declivity, slope exposure, depth of landform fragmentation and the hydrographic network density. The techniques used in creating the land declivity and slope exposure maps implied spatial modeling in TNTMips (Microimage). The maps for hydrographic network density and depth of landform fragmentation were created in ArcGis by applying the functions of the density for lines per surface units (1 km^2) and by calculating the lowest and highest altitudes with the help of FocalMin and FocalMax functions.

In order to create the distribution graphs and the histograms for the frequencies of statistical data, they were exported from the databases of vector layers and from the histograms of raster layers by theme, the last being processed in OpenOffice (http://www.openoffice.org). The classification of the vector layers and raster by theme played an important role in creating the graphs and histograms. In the case of vector layers by theme, the classification implied making internal tables with a one-to-many architecture. A recording is attached to several vector elements: for example, in the table, the hierarchic order of the hydrographic network with the value 1 was registered once and then attached to several river lines or segments, or polygons, thus facilitating database interrogation on quality categories and data extraction only for those categories.

In order to clearly underline the frequencies for certain parameters (histogram for the frequencies regarding the depth of landform fragmentation classes), the graph was presented in a logarithmic scale. The statistical data of the raster maps were extracted from the corresponding histograms, imported in OpenOffice, classified and processed. For certain themed layers (land declivity, slope exposure), the histograms of class values were extracted after being classified depending on the technique of spatial manipulation language (SML).

The Crest of Ghilcoş-Hăghimaş-Mezinul Mountains is included in the group of average mountains. The highest altitude is 1,786 m, in the Hăghimaşul Mare Peak. The crest is very fragmented in the upper side. The unitary aspect of the main crest is accentuated by the existence of the deep Bicaz valley and of the depressions flanking it (Bojoi 1964).

The morphometry is the result of a long evolution process of the hydrographic basin influenced by the action of water. This is the reason why there are interdependence relationships between the morphometric parameters and the (liquid and solid) discharge. The most important morphometric parameters related to the hydrologic characteristics are the following: surface; altitude; average slope of the hydrographic basin; length and slope of the mainstream, etc.

The surface of the catchment basin is 40.59 km^2, relatively small compared to the size of the lake. This very aspect has facilitated a low-clogging ratio and the endurance of the lake. The surface of the hydrographic basin is drained by a network of water streams with sizes depending on those of the drained surfaces. The Red Lake hydrographic basin has developed in a relatively uniform manner on both sectors, with relatively low dominance on the left.

Surface is the main element directly relating the water flows. In this case, the basins with similar physical-geographic conditions, with similar sizes, may have flows of about the same value. The maximum length, on the NW–SE direction, is 10.78 km and the average 3.76 km. The total length of the drainage network is 166 km with most arteries having an intermittent character (Fig. 5.3).

The maximum width of 5,900 m is found in the northern side of the basin. The elongated shape does not allow the accumulation of a large amount of water; thus,

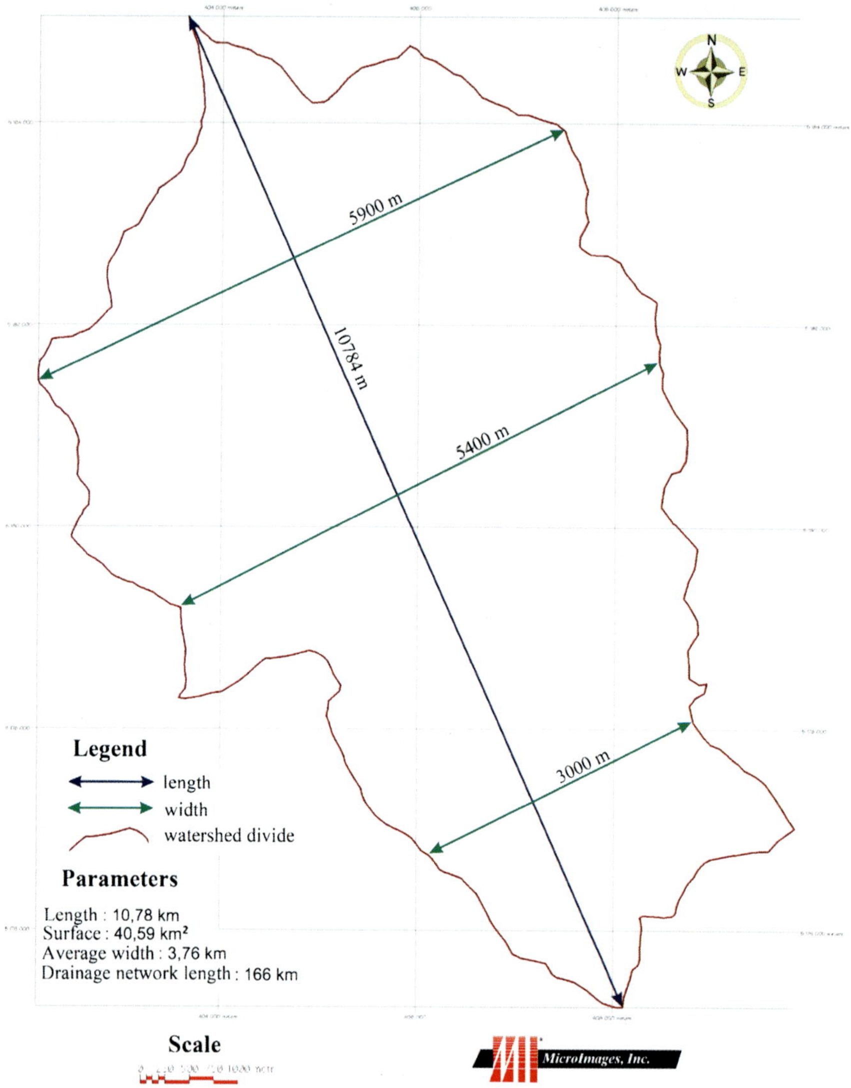

Fig. 5.3 Morphometric characteristics of the Red lake catchment basin

catastrophic floods are rare. The watershed can reach altitudes exceeding 1,500–1,600 m. The water surface is situated between altitudes of 966 and 965 m.

The most important water streams are the following: Suhard, Licoş, Vereschiu, Oaia (Oii), Calul, Sec, Zgomotos, Hăghimaş and Hăghimaşul Negru. All these brooks have a permanent character. The other arteries have intermittent streams, most of them having water only during heavy rains.

The most important brook is Oaia (Oii) (the former upper stream of Bicaz), supplying the lake in the southern sector, followed by the Vereschiu and Suhard brooks (Fig. 5.4). They are also the brooks opening directly onto the lacustrine basin, the others being only direct and indirect tributaries. The diverse physical-geographic conditions make the determining elements in defining the plan-specific configuration of the hydrographic network insufficient for the genetic basis of river network classification.

The first attempt to classify rivers starting from the positions of streams in relation to the main collector was elaborated by Gravelius (1914, cited by Horton 1945). According to this system, the largest river is considered to be of first order from source to mouth. In time, the ordering changed (Horton 1945), attributing the first order to the smallest fingertip tributaries appearing as a result of concentrated runoff, with no other tributaries. A second-order stream is one that receives at least one or several tributaries of the first order and only of this order. If a second-order stream joins another of the same order, a third-order stream appears and so on. This happens until the highest order in a hydrographic basin is reached. This way, a stream maintains its order from source to mouth.

According to the Horton-Strahler classification system, the tributaries of the Red Lake reach the sixth-order. Most of them are of first-order, summing up to 146.75 km (50.44 %), followed by the second-order streams with 33.78 km (21.85 %), the third-order streams with 21.13 km (12.29 %), the fourth-order streams with 13.01 km (9.85 %), the fifth-order ones with 6.01 km (4.8 %) and the sixth-order ones with 1.24 km (0.88 %) (Fig. 5.5).

Given the high landform fragmentation, the slopes have different orientations. From this viewpoint, the plain-like surfaces occupy 0.4 %; those with a NE orientation—16.5 %; E—14.1 %; N—13.9 %; NW—13.5 %; SW—11 %; W—10.9 %; S—9.9 %; and SE—9.7 % (Fig. 5.6). There are numerous slopes with a northward orientation (N, NE, and NW), with a 43.9 % percentage, leading to increased humidity and lower evapotranspiration. In the northern part of the basin, because of the hydrographic network location, with a west–eastward orientation, there is a low frequency of northward orientation and a high frequency of southward orientation. The eastern sector of the slope within the Oaia-Bicaz main valley has a dominant westward orientation. It should be underlined the very low frequency of plain-like surfaces, with a 0.4 % percentage.

Given the fact that the Red Lake has a north–south alignment and that there is a high surrounding landform, there is different luminosity on the two lateral slopes—eastern and western—depending on the height of the surrounding mountain units. The surface does not receive the same quantity of energy as the surrounding areas

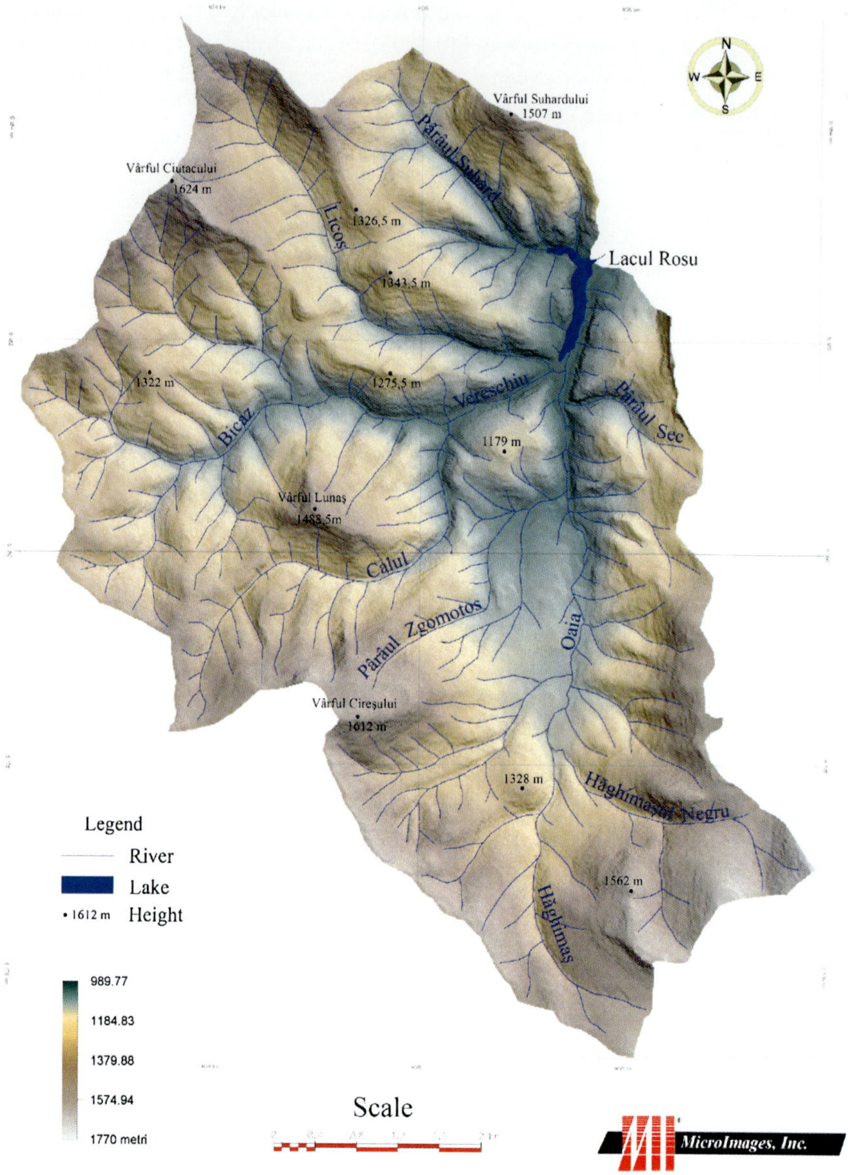

Fig. 5.4 Map of the drainage network and morphology of the Red lake hydrographic basin

as the sunrays touch the lacustrine surface much later at sunrise and then they go down late, around sunset.

A special influence is one of creating topoclimates in the Red Lake area. Hence, three different sectors are created, depending on the orientation of the main hydrographic arteries opening onto the water and on the orientation of the

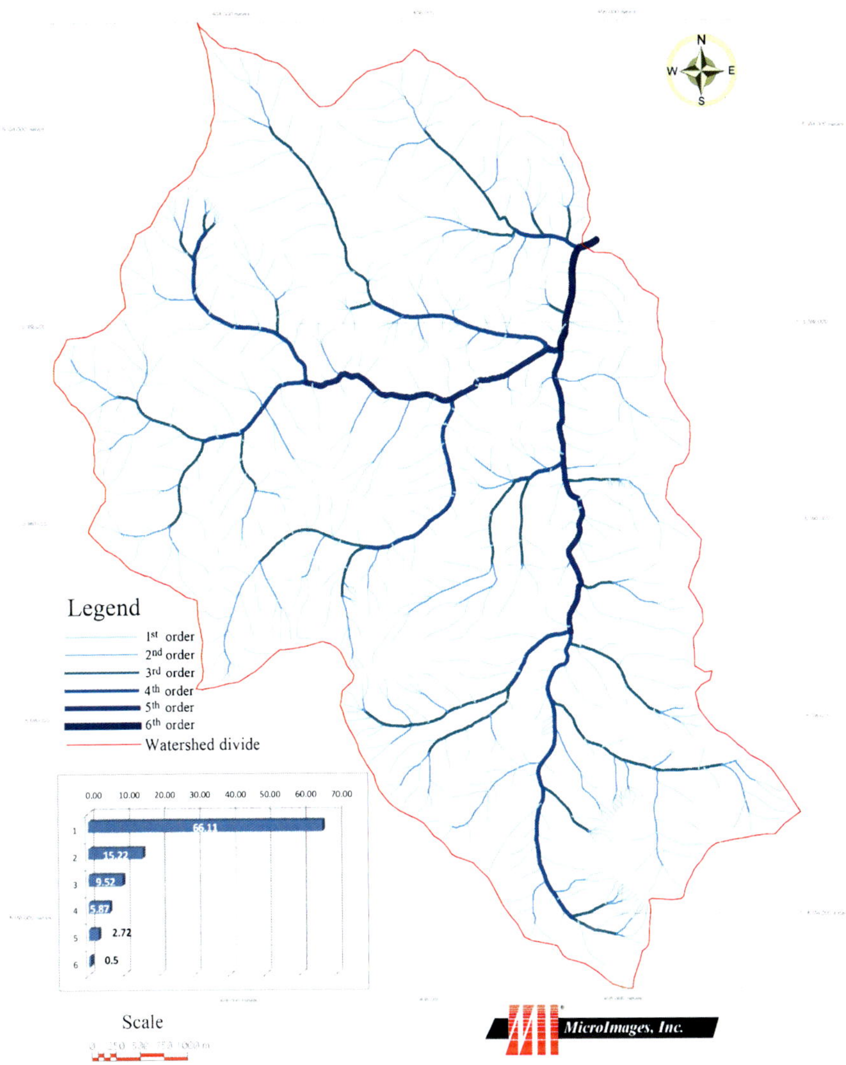

Fig. 5.5 Map illustrating the hierarchy of the hydrographic network according to the Horton-Strahler system

surrounding slopes. The Suhard artery has a west-eastward general orientation, hence a better illumination of the waters but a lower illumination for the slopes with a northern or southward orientation. The artery of the Oaia (Oii) brook, with a south-northward orientation, will have a lower illumination of the waters but a better one for the slopes, with an eastern or western dominance. There are uniform proportions only for the confluence sector.

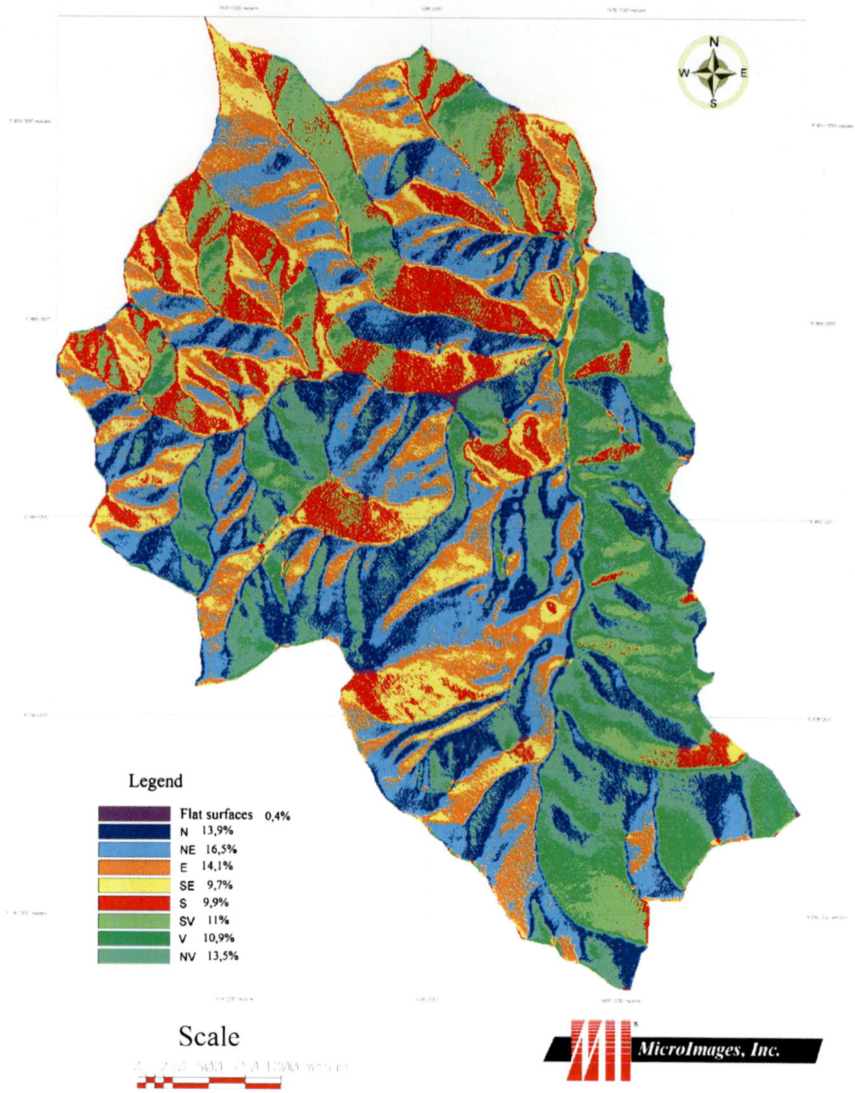

Fig. 5.6 Map of the slope exposure in the Red lake hydrographic basin

The first-order corresponds to 517 river segments and a total length of 146.75 km. The sixth-order corresponds only to 9 river segments and a total length of 1.249 km (Table 5.3, Fig. 5.7).

The relative frequency of the drainage networks, of I-VI orders, clearly demonstrates the dominance of the first-order basins (50.44 %) and the low proportion of the sixth-order basins (0.88 %) (Fig. 5.8). There are 517 first-order hydrographic basins, with a total surface of 29.176 and 11.415 km^2 of interbasinal

Table 5.3 Relation between the network order and the number of river segments

Drainage network order (Horton-Stralher)	No. of river segments	Relative frequency (%) total = 1,025 river segments	Length (km)
I	517	50.44	146.75
II	224	21.85	33.784
III	126	12.29	21.136
IV	101	9.85	13.019
V	48	4.68	6.015
VI	9	0.88	1.249

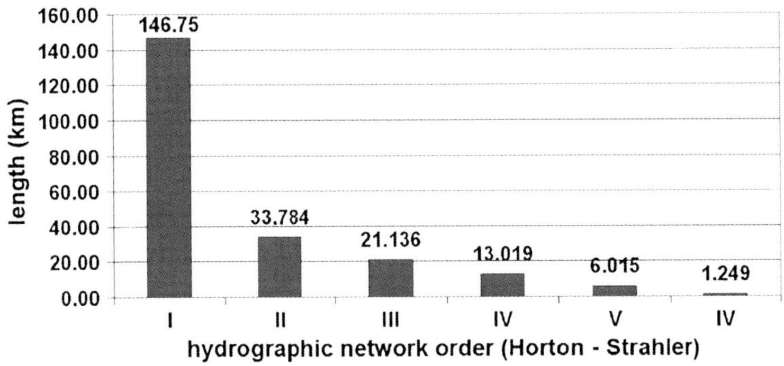

Fig. 5.7 Cumulative length depending on the drainage network orders circumscribed to the Red lake hydrographic basin

surfaces (Table 5.4). The relative frequency of the hydrographic basins, of I–VI orders, clearly demonstrates the dominance of the first-order basins (78.45 %) and the low proportion of the sixth-order basins (0.15 %) (Fig. 5.9). The cumulative surface of the first-order hydrographic basins is 29.176 km^2 and that of the sixth-order basins is 40.591 km^2 (Fig. 5.10).

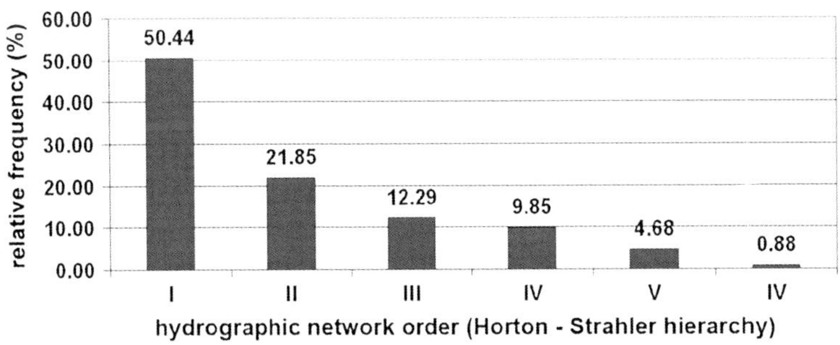

Fig. 5.8 Histogram illustrating the relative frequencies of the drainage network—*I–VI* orders circumscribed to the Red lake hydrographic basin

Table 5.4 Classification of hydrographic basins and the surfaces corresponding to each order

Hydrographic basin (Horton-Stralher)	No. of hydrographic basins	Relative frequency (%) total = 659 basins	Surface (km^2)	Interbasinal surface
I	517	78.45	29.176	11.415
II	103	15.63	20.993	19.599
III	28	4.25	24.651	15.940
IV	8	1.21	32.870	7.721
V	2	0.30	35.167	5.424
VI	1	0.15	40.591	0.000

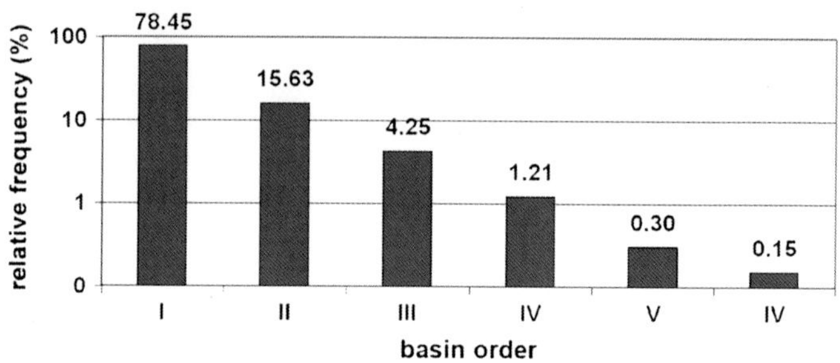

Fig. 5.9 Histogram illustrating the relative frequencies of the hydrographic basins—*I–VI* orders circumscribed to the Red lake hydrographic basin

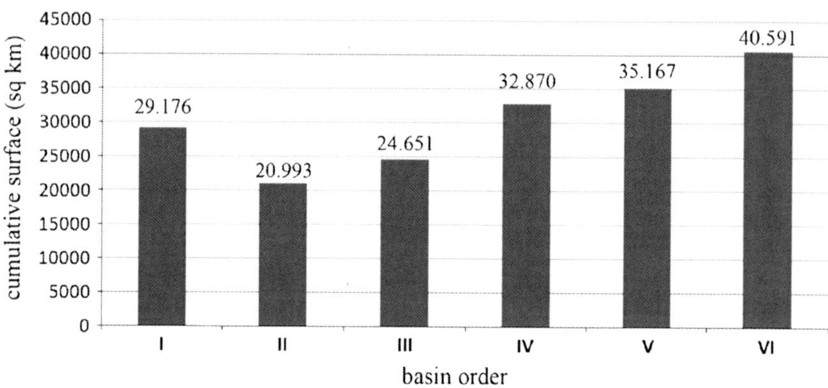

Fig. 5.10 Cumulative surface by hydrographic basin orders circumscribed to the Red lake hydrographic basin

Because of the mountain morphology, the first-order hydrographic basins have extremely diverse shapes, most of them being elongated (Figs. 5.11, 5.12, 5.13). The interbasinal surfaces represent a small proportion.

The second-order hydrographic basins are located, most of the times, in the upstream sectors of the main streams (Figs. 5.14, 5.15, 5.16). The third-order

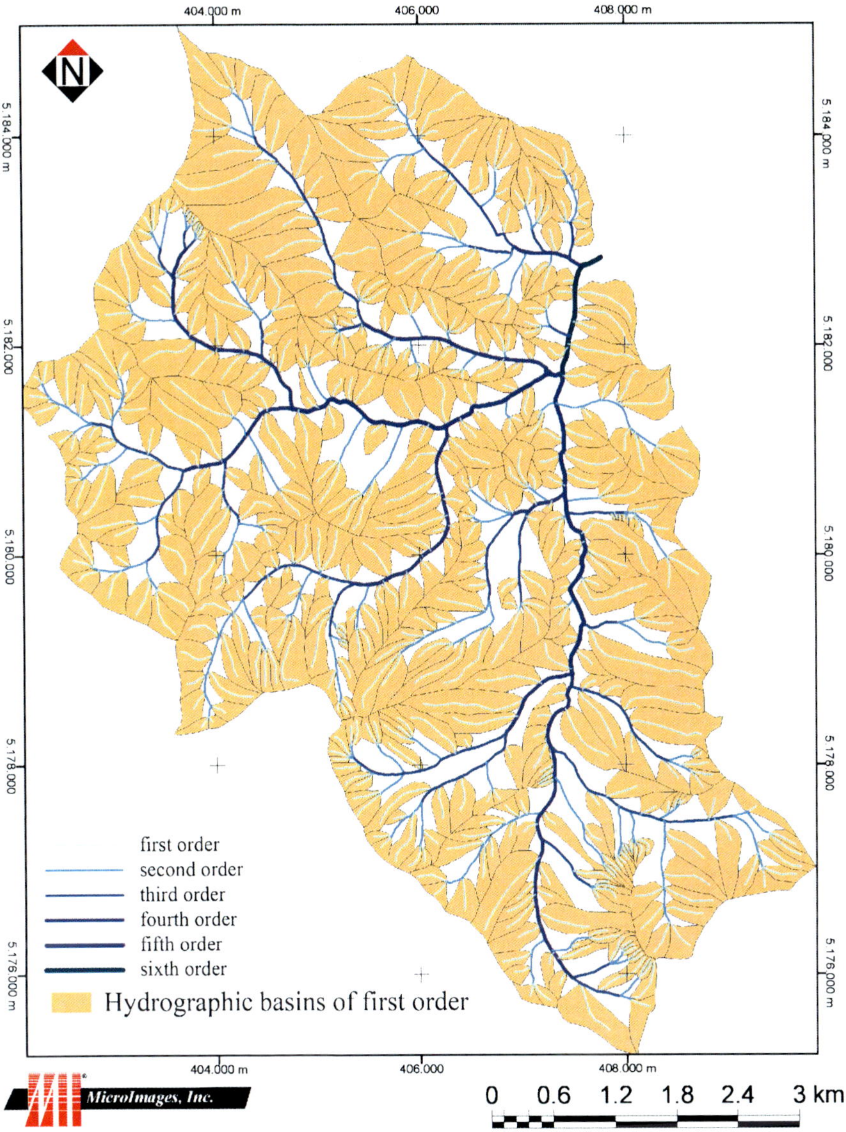

Fig. 5.11 Emphasis on the first-order hydrographic basins

hydrographic basins occupy the upstream half of the main streams. Most of the times they have triangular shape and they are located in the upper half of the hydrographic basin (Figs. 5.17, 5.18, 5.19).

The fourth-order hydrographic basins occupy almost the entire basin surface of the Red Lake. They are very elongated and the water flows are rather important, being active almost year round (Fig. 5.20). Only the downstream extremity of the

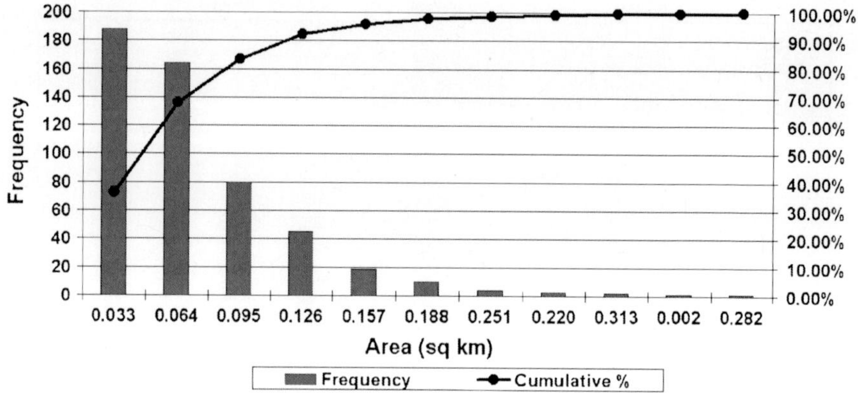

Fig. 5.12 Histogram of raw frequencies and of the *cumulative percentage curve* for the surfaces of first-order hydrographic basins (according to the Horton-Stralher classification of the drainage network, extracted from the 1:25,000 DTM maps 1982)

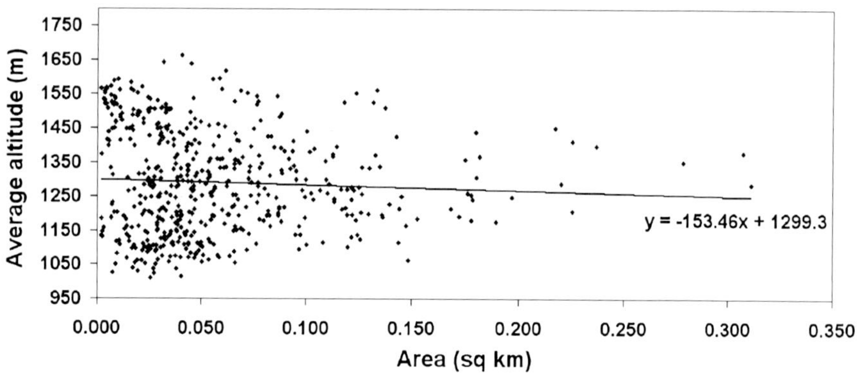

Fig. 5.13 Graph illustrating the correlation between the surface of first-order hydrographic basins and their average altitude

Red Lake is an exception, not being part of this category. There are two-fifth-order arteries, joining the Red Lake basin downstream (Fig. 5.21).

The Red Lake basin, on the Oaia (Oii) valley alignment, is represented by a sixth-order stream (Fig. 5.22). The drainage density is an indicator of the landform evolution and fragmentation of surface runoff, of the rainfall degree, as well as of flows and of alluviums within a hydrographic basin.

The comparative analysis between the density of the hydrographic network (permanent streams) and the drainage density (the entire network of valleys) underlines extreme quantitative differences. Functionally, both represent the drainage capacity of the soil. The quantitative difference is the fact that, all the channels of surface drainage were taken into account for the drainage density, regardless of their permanent or temporary character, while for the density of the

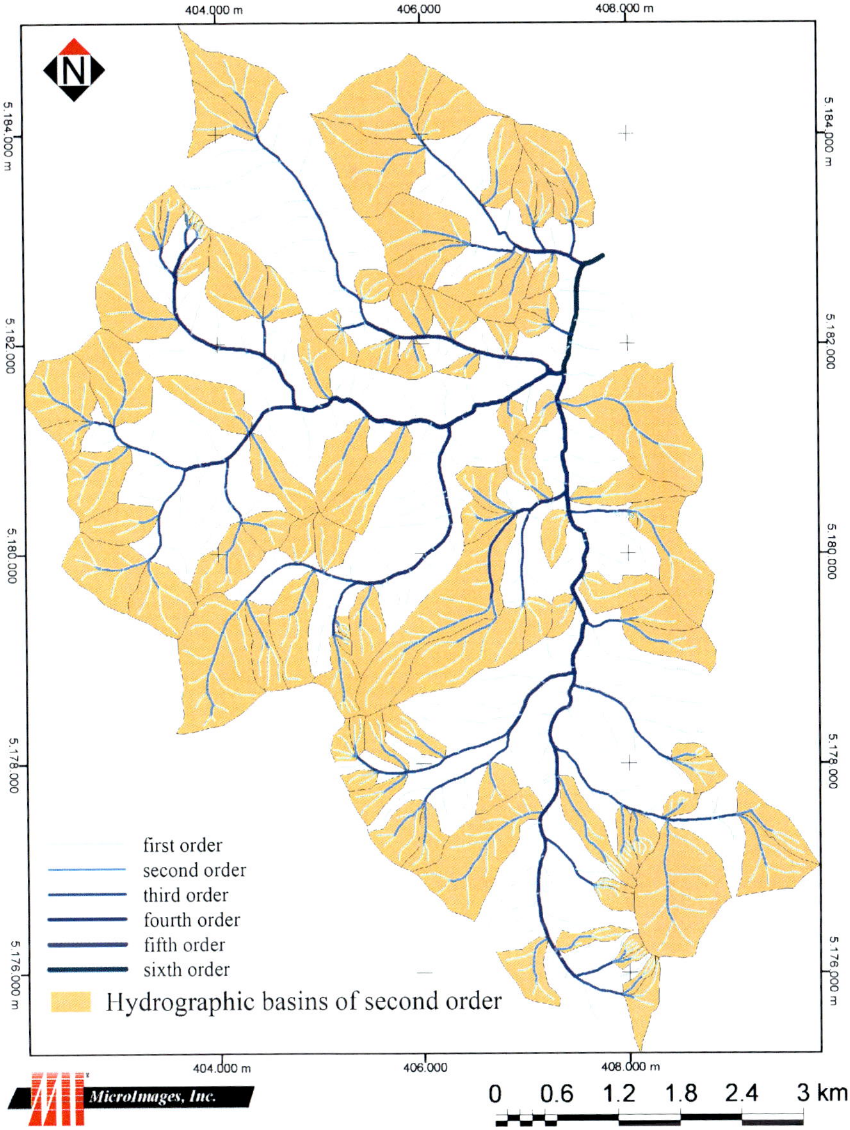

Fig. 5.14 Emphasis on the second-order hydrographic basins

hydrographic network only the length of permanent or semi-permanent drainage channels was calculated. In this case, the density of the hydrographic network is the same as the density of permanent drainage. Drainage density implies the length of all permanent and temporary runoff channels.

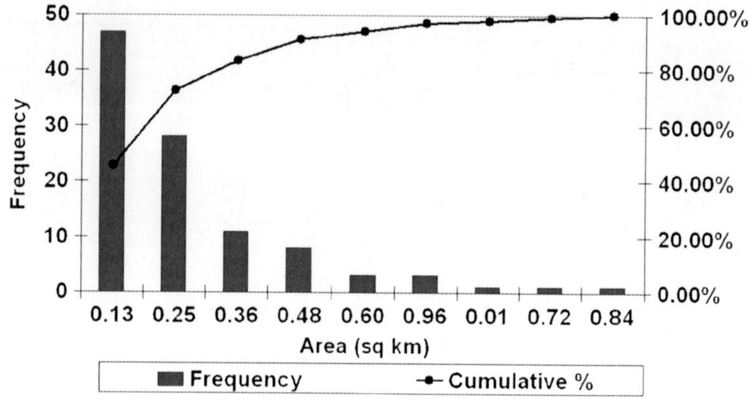

Fig. 5.15 Histogram of raw frequencies and of the *cumulative percentage curve* for the surfaces of second-order hydrographic basins (according to the Horton-Stralher classification of the drainage network, extracted from the 1:25,000 DTM maps 1982)

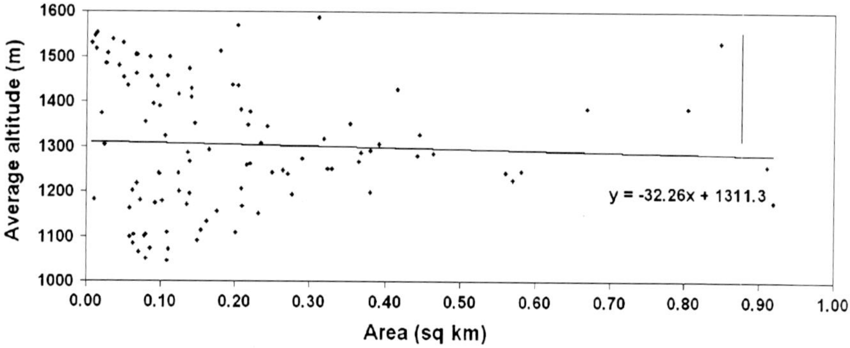

Fig. 5.16 Graph illustrating the correlation between the surface of second-order hydrographic basins and their average altitude

The highest values of landform fragmentation depth correspond to the mountain units dominated by friable rocks, such as limestones and dolomites. The lowest values belong to depressions, valley corridors and structural plateaus (Figs. 5.23, 5.24). The highest values of landform fragmentation depth, of over 400 m, are registered in the north-eastern sector (Suhardul Mic), with calcareous rocks and abrupt landforms such as saddles, as well as between the upper Bicaz and the Calului brook, dominated by dolomites and deep valleys, with a sharp "V"-shape profile (Fig. 5.25).

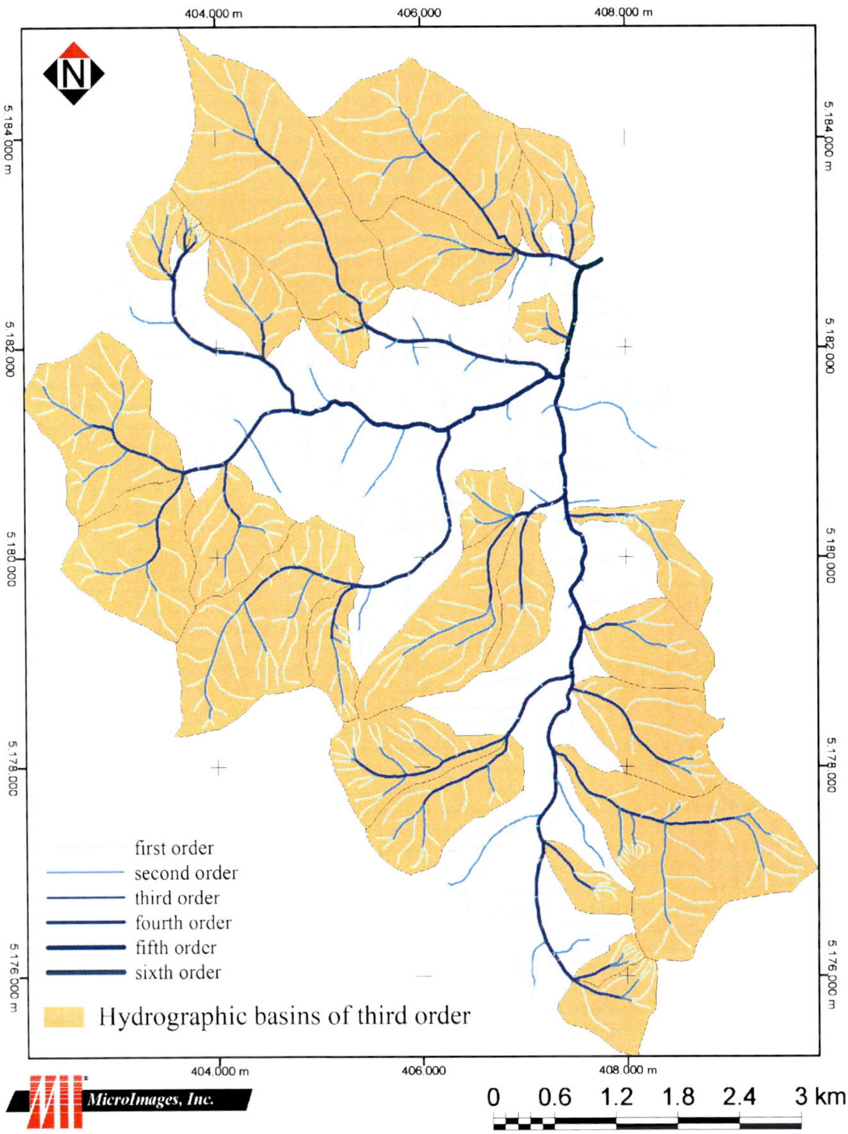

Fig. 5.17 Emphasis on the third-order hydrographic basins

The highest values also correspond to landforms with friable rocks (limestone, dolomites), easy to model. The lowest densities correspond to structural plateaus and to crystalline rocks, which are harder and not easy to model (Figs. 5.26 and 4.3). The density class of the hydrographic network, with the highest frequencies, representing 25.75 %, is between 5.5 and 6 km/km^2 and the lowest, of 0.003 %, between 0.5 and 1 km/km^2 (Fig. 5.27).

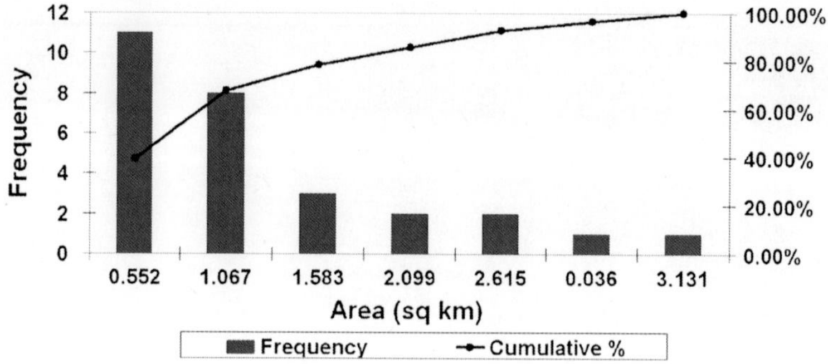

Fig. 5.18 Histogram of raw frequencies and of the *cumulative percentage curve* for the surfaces of third-order hydrographic basins (according to the Horton-Stralher classification of the drainage network, extracted from the 1:25,000 DTM maps 1982)

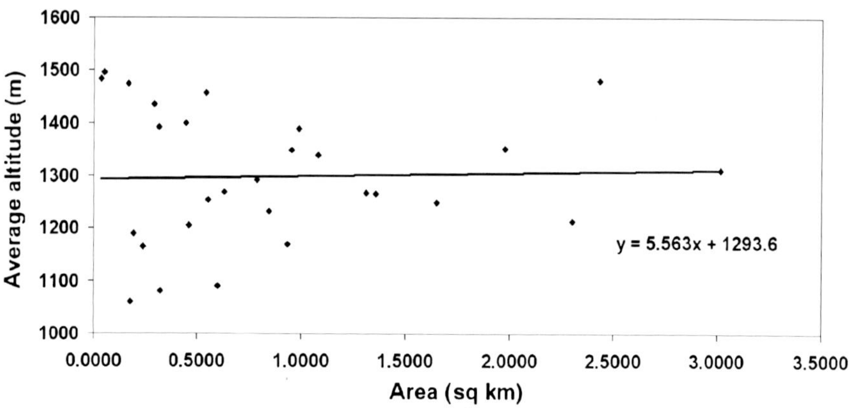

Fig. 5.19 Graph illustrating the correlation between the surface of third-order hydrographic basins and their average altitude

5.3 The Red Lake Lacustrine Basin

The field measurements were taken by using the LEICA TCR 1201 total station, which, together with LEICA GPS 1200, is part of SYSTEM 1200 LEICA. The Topographic Headquarters in Gheorgheni provided us the coordinates of the topographical marks. The measurements were taken starting from Topographic Mark number 15, located on the right side of the lake, at 986.275 m altitude. Orientation was done towards the mark on the Suhardului peak, at 1,507.0 m altitude. After finishing the measurements, the data were processed by AutoCAD software. At the same time, a GPS was used in order to establish the exact perimeters of the lacustrine basin and the typical wetlands around the Red Lake.

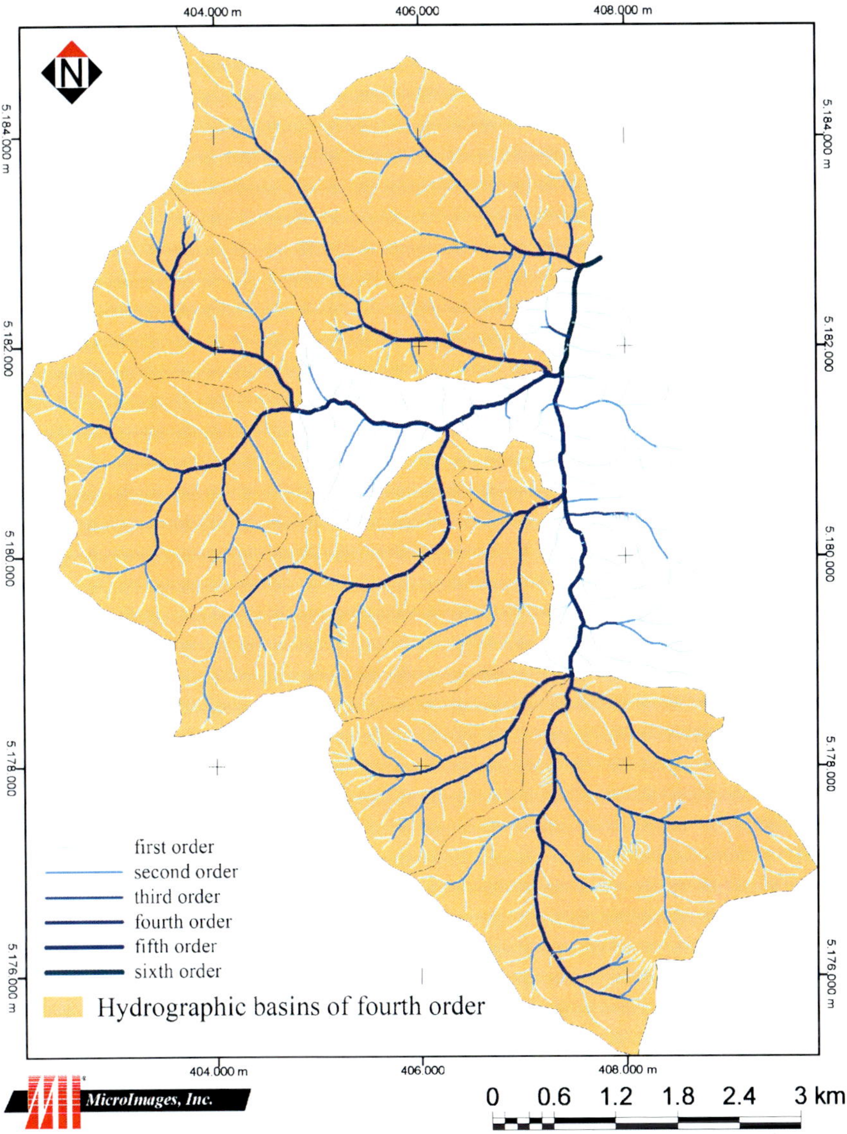

Fig. 5.20 Emphasis on the fourth-order hydrographic basins

The data of the bathymetric surveys were represented in Stereographic Projection 70. All the topo-geodesic works carried out on the territory of Romania were executed in Stereo Projection System 70 or Stereographic 1970: maps and cadastral plans, topographical maps, etc. The bathymetric measurements were taken by using a Valeport Midas Surveyor echo sounder (Bathy-500DF Dual

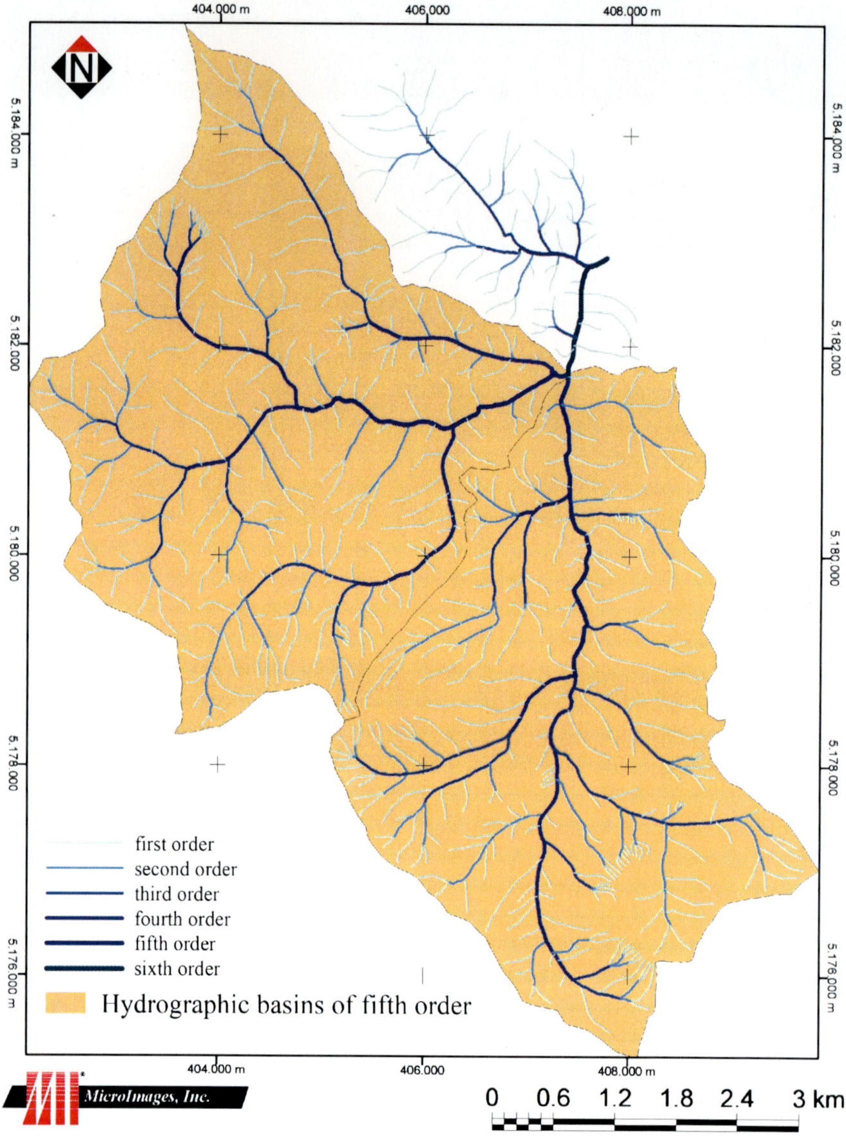

Fig. 5.21 Emphasis on the fifth-order hydrographic basins

Frequency Hydrographic Echo Sounder). The resolution of this echo sounder is 1 cm/1 cm. It includes GPS navigation.

The contour of the lake was established based on topographical measurements and GPS. The entire lacustrine surface was measured by using a sonar. In order to process the bathymetric data and create the thematic maps, TNTMips v.7.2 and ArcGis v.9.3 software were used. This way, the numerical model of the land was

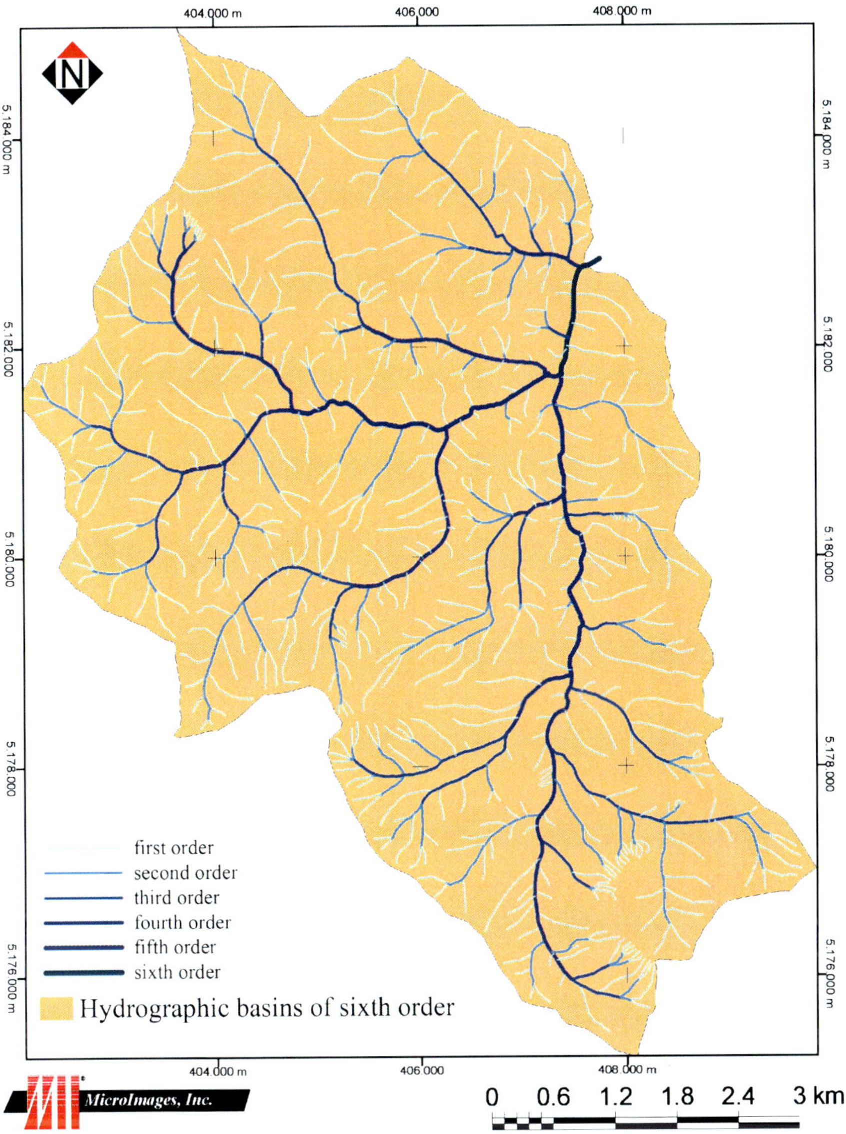

Fig. 5.22 Emphasis on the sixth-order hydrographic basins

created, which was used graphically for different purposes. These are novel graphical representations for the mathematical modeling of lacustrine basins.

The silt samples were taken from the bottom of the lake by means of a goblet. The total organic carbon was analyzed in the laboratory according to the Walkey-Black formula (Baena et al. 2007). By multiplying the organic carbon by a factor of 1.724, the value of the humus in the bottom lake deposits was established

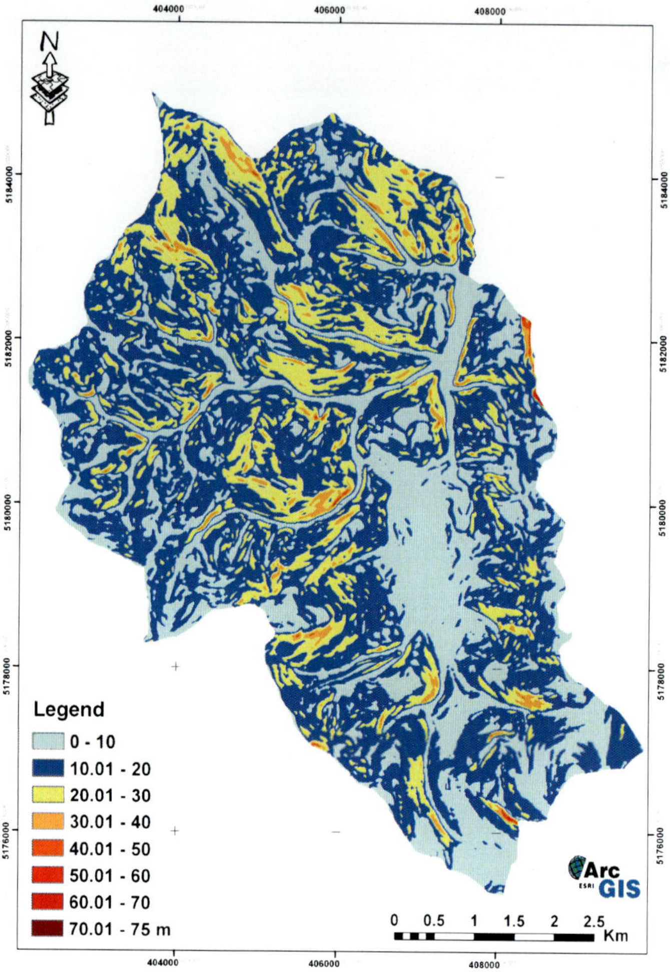

Fig. 5.23 Map of landform fragmentation depth for the Red lake hydrographic basin

(Walkley and Black 1934). The physical–chemical water parameters were measured by means of a Hach multiparameter device (manufactured in Germany). The identification of species was made based on the recognition of a surface of 10 × 10 m (Nechita 2003).

5.3.1 Morpho-Bathymetric Parameters

The topographic measurements indicated the following values: area—12.01 ha (120,134.44 m^2); perimeter 2,905.79 m; maximum length—1,361.72 m; maximum width—160.24 m; maximum altitude upstream in the south (the outlet of the

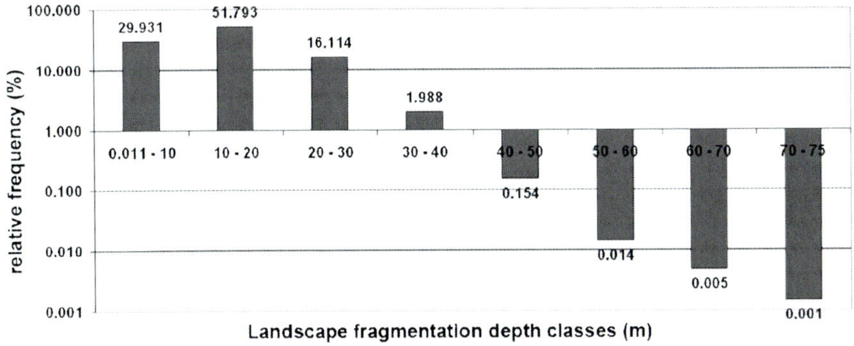

Fig. 5.24 Histogram illustrating the relative frequencies of the depth of landform fragmentation classes for the Red lake hydrographic basin

Fig. 5.25 Map of the depth of landform fragmentation according to the method of squares

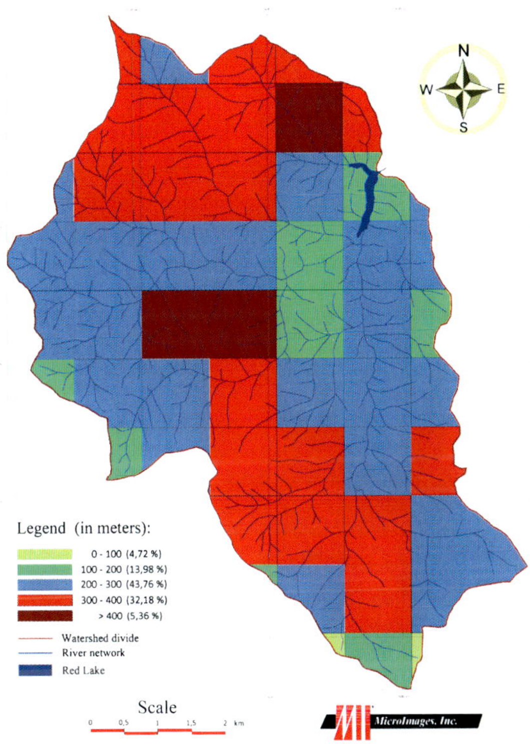

Oii stream) 965.199 m; maximum altitude upstream in the north–west (the outlet of the Suhard stream) 966.041 m. The length of the main sector (the outlet of the Oii stream and the river mouth) is 958.89 m. The length of the secondary sector (the outlet of the Suhard stream and the river mouth) is 403.32 m (Fig. 5.28). The

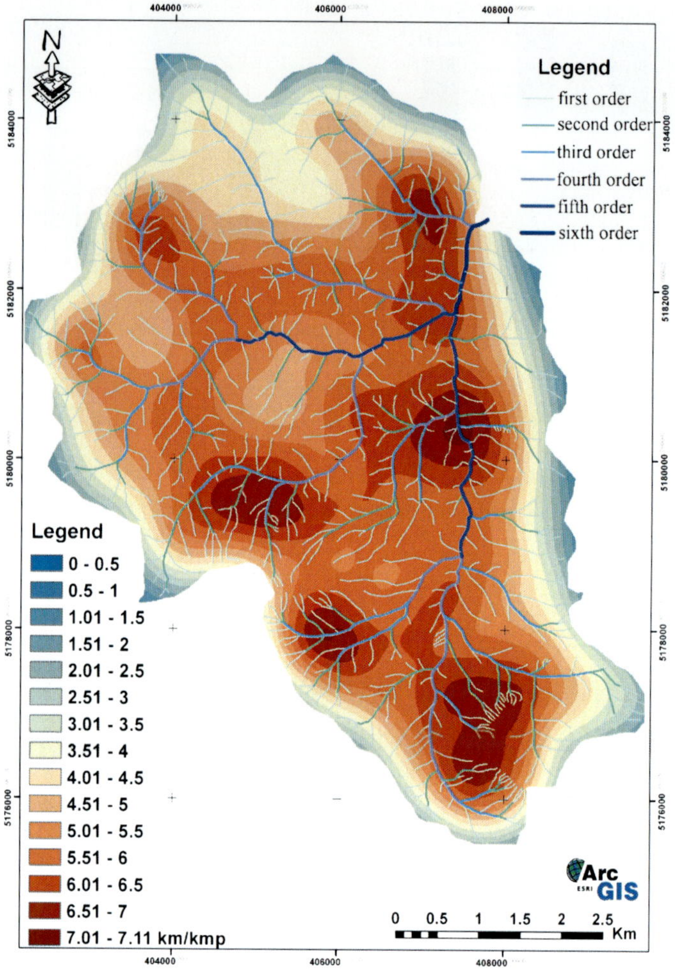

Fig. 5.26 Map illustrating the density of the hydrographic network for the Red lake hydrographic basin

area indicated in other recent studies is 11.6 ha or 116,500 m^2 (Gâstescu 1971, 1979, 1998; Pandi 2004).

Initially, the lake was 2 km long on the Oaia (Oii) stream and 1.6 km long on the Suhard stream (Dragoş 1957). In 1957, the axial length on the Oaia (Oii) stream was 1 km and of the Suhard stream, 442 m. The total surface was 12.6 km, with a 5.4 average depth and a 10.5 m maximum depth (Pişota and Năstase 1957) (Fig. 5.29).

The river mouth, or the tributary sector of the Red Lake, is located at an altitude of 965.05 m. The difference of level between the outlet of the Oaia (Oii) stream and the river mouth is 14 cm. There is a level difference of 99 cm between the

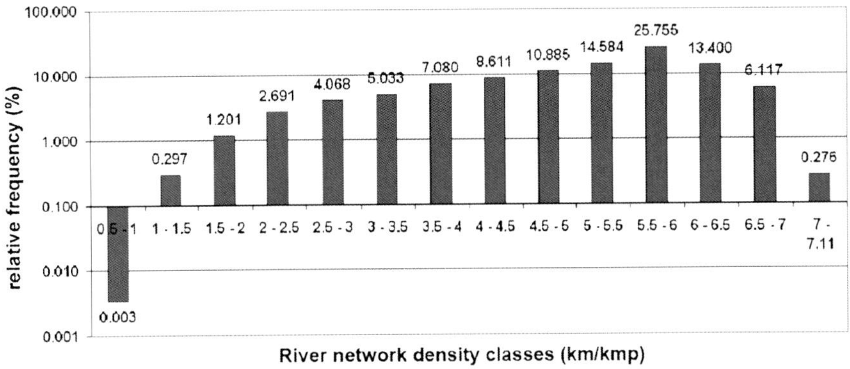

Fig. 5.27 Histogram illustrating the relative frequencies of drainage network density classes for the Red lake hydrographic basin

outlet of the Suhard stream and the river mouth. The slope of the Suhard stream is therefore greater and the bottom of the valley is deeper. The smaller distance also indicates a greater relative relief. The Oaia (Oii) stream and the adjacent lacustrine sector, implicitly, have a greater length but a smaller slope. In this case, the sedimentation rate could be higher, as the current has a lower speed and the elimination of the alluvia is slower.

There is a relatively great difference between the altimetric data of the lacustrine basin, especially for the average level of the lake, in the studies of different authors: Pişotă and Năstase 1957; Udriste 1963. According to the data obtained in the measurements taken in 2009, the altitude of the lake level is 965.10 m but in other recent studies, a value of 978.17 m is indicated (Pandi 2004).

The difference could be explained by the different methods that were used in establishing this altitude (Clague et al. 2006). In the present case, topographic measurements were taken, while in other studies, the altitude was established based on available topographic maps.

The most reduced depths are registered in the upstream sectors of the two branches, at the outlet of the most important streams: Oaia (Oii) and Suhard. This fact is a result of the relatively reduced transport of solids from the two water streams (Fig. 5.30). The reduced discharge of solids is the result of a dense forest cover maintained since the formation of the lake. Reduced depths were also registered next to the river mouths of the small torrents coming from the mountain sectors with high slopes.

The greatest depth (10.5 m) was registered in the central sector, at the confluence of the two main inflows or behind the landslide dam. The higher level difference between the Suhard stream and the river mouth causes the appearance of greater depths along this profile, as compared to the one located between the Oii stream and the outlet. Until now, only maximum depths of 9.70 m have been measured (Pandi 2004) (Fig. 5.31).

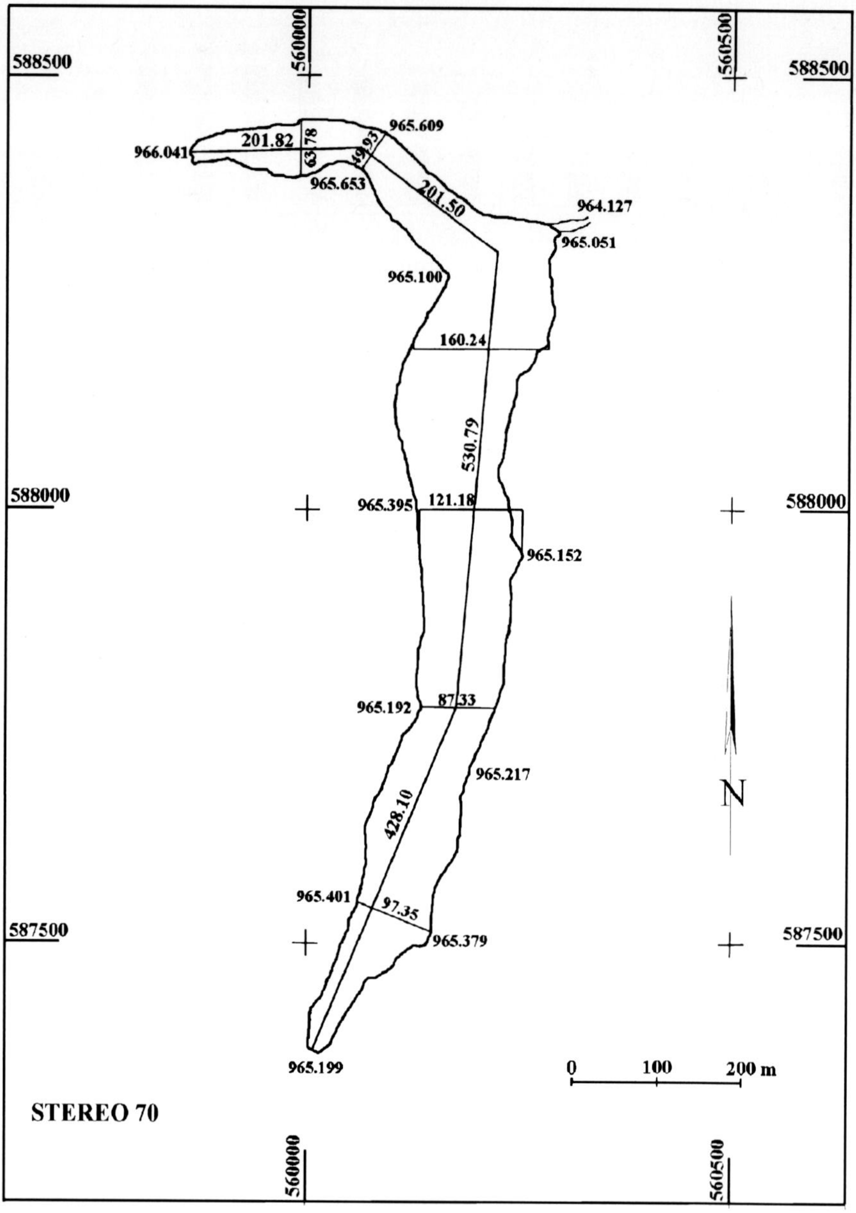

Fig. 5.28 The topographical survey of the Red lake bank line in 2009 (stereo 70 projection)

The measurements taken in the summer of 2009 indicated values of approximately 10.50 m. Only one study indicates a depth of 10.5 m (Cazacu et al. 2009). It is important to mention the fact that these measurements were taken in a period when major amounts of precipitation were registered over a long period. This fact

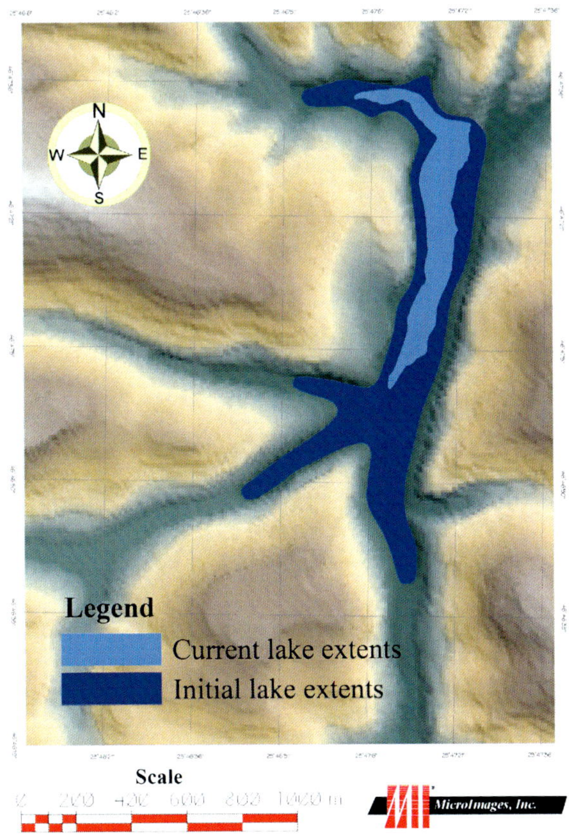

Fig. 5.29 Initial and current surface of the Red lake

caused a slight increase in the water level but not higher than 20–30 cm. From this point of view, there are no accurate figures on the periodic variations of the levels, as the hydrologic observations are not made systematically and in most cases, they are false. All measurements were carried out in real time (Fig. 5.32).

In a transverse profile, a clear asymmetry of the slopes can be noticed. In the remote sector, towards Ghilcos Mountain, from where the landslide material was detached, the slope becomes gentler, while the opposite slope is very steep. The highest relief is registered in the sector of the river mouth, where the greatest depths are also found. A perfect symmetry was registered in the upstream sectors, where the silting is predominant and the quantity of the landslide material is reduced (Fig. 5.33).

The steepest inclinations were found on the western and northern banks, where the rocky slopes are almost vertical above the lacustrine basin. These sectors are well forested. They do not represent the source for any significant water streams and the quantity of alluvia material is low. A different situation occurs on the opposite (eastern) slope, with landslide diluvium. In this case, the slopes have reduced values and plant communities could develop. This latter sector is strongly

Fig. 5.30 Connection between the Oaia (*Oii*) (*left*) and Suhard (*background*) arteries, upstream from the mouth

Fig. 5.31 Area of the Red lake where the highest values of depths are registered

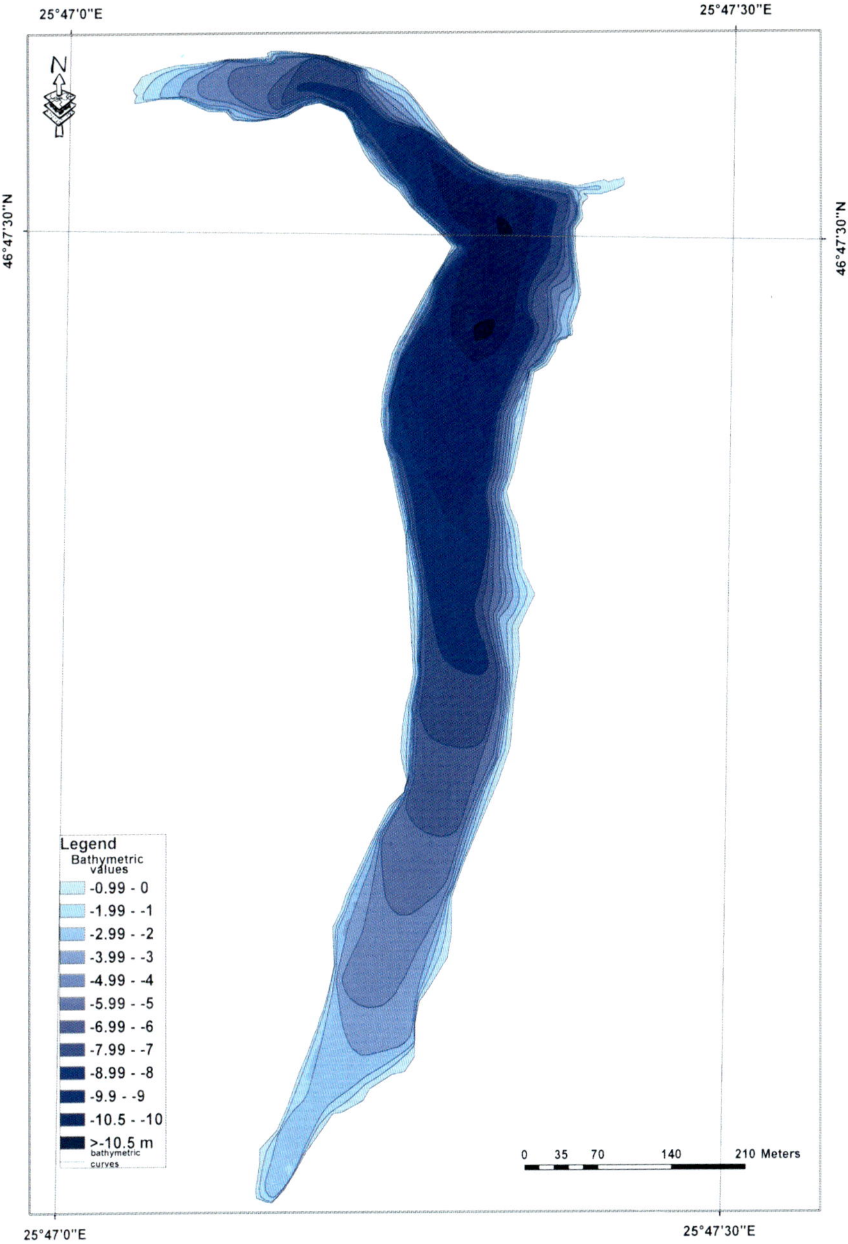

Fig. 5.32 Bathymetry of the Red lake lacustrine basin

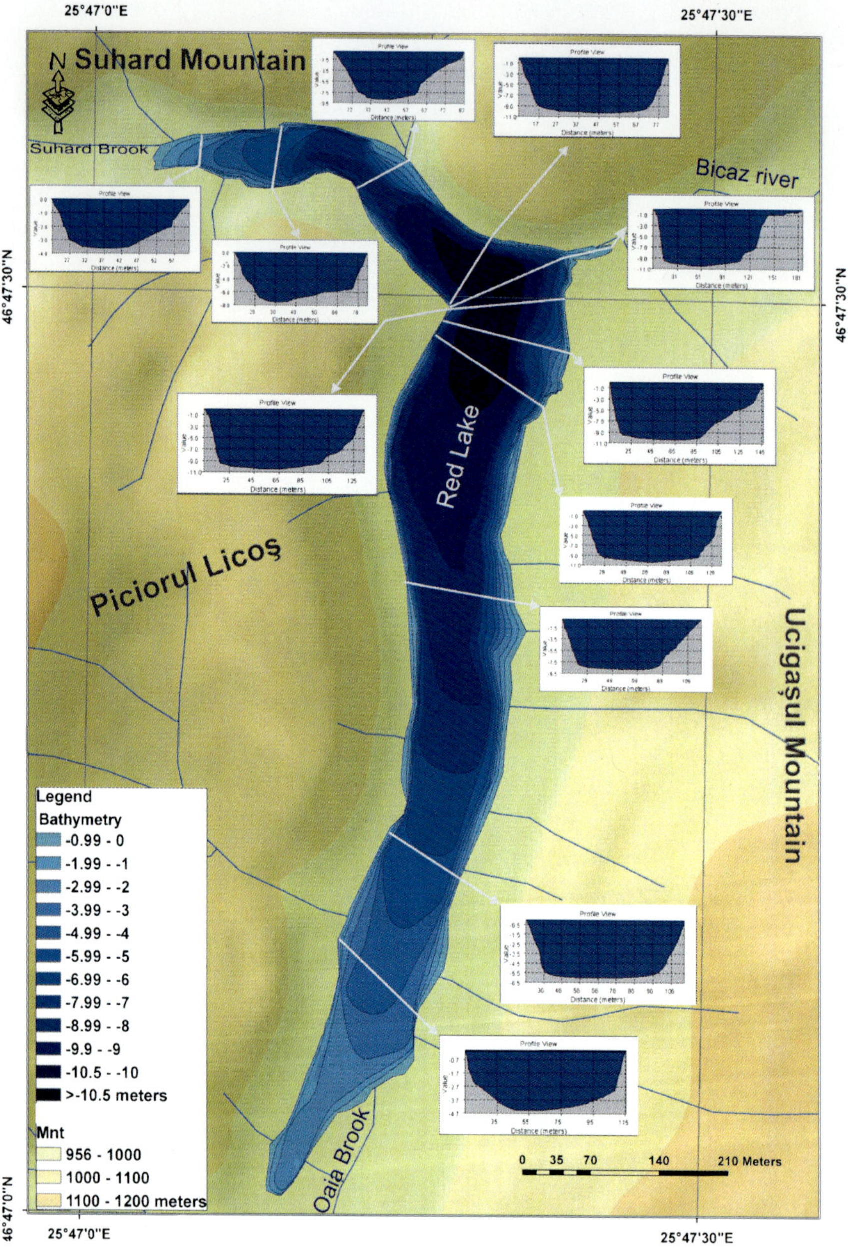

Fig. 5.33 Transversal profiles in the Red lake lacustrine basin

fragmented by several torrents. They are anthropically controlled at present, as the linking road passes this area between Bicaz and Gheorgheni.

It should be mentioned that in the areas with reduced depths, where light penetrates to the bottom of the lake, a vegetation specific to the extensive wetlands develops. Vegetation significantly contributes to the rapid silting in the upstream sectors of the two branches. At the same time, it created a soil typical of the lacustrine basins, called limnic soil (Islam 2010; Mfundisi 2008; Mujdeci et al. 2010; Romanescu and Cojocaru 2010; Romanescu et al. 2010a, b).

Bathymetry plays the role of altitude. In this case, the hydro-bathymetric conditions are layered, just as the climatic layering within the higher landforms. The role of bathymetry in interpreting the hydrologic phenomena is extremely complex (Gagnon et al. 2008). For a better understanding of the phenomenon, a series of maps underlying the phenomena studied, separate or complex, are used (Godet et al. 2009).

The letter "L" shape is due to two major valleys: Oaia (Oii) in the central-southern half and Suhard in the northwestern sector. The landslide body, with a clear asymmetric transverse profile, affects the central-southern branch of the Oii valley. Landslides do not affect the branch of the Suhard valley but it has an asymmetric transverse profile due to significant silting towards the northern bank. From this bank, several torrential water-streams deposit material in small submerged cones of fan-delta type. The same micro-morphological formations can also be found on the eastern slope of the main valley.

The water level is situated at 965–966 m and definitely not at an altitude of 978 m as indicated by Pandi in (2004). In fact, the whole range of morphometric data should be revised, as, until the present date, it is only the information taken from the topographical maps drawn between the 1950s and 1970s that has been used.

Slopes between 0 and 3° are most common (over 50 %) and the lowest occurrence values are found for slopes between 45 and 66° (under 2 %) (Fig. 5.34). A quite remarkable percentage (almost 10 %) is registered for slopes between 5–10° and 15–25°. At reduced depths (0–1, 1–2 m) the average slopes are predominant, while at greater depths (especially for the depths between 8–9, 9–10, over 10 m), the reduced slopes are predominant (Fig. 5.33). For the depths exceeding 9 m the slopes with values between 0 and 3° are predominant. Therefore, a uniformity of the lake bottom is observed.

The high frequency of low-value slopes also indicates the relative high degree of clogging. Rocky walls give the bigger slopes, mostly by those within the western and northern sectors of the basin (Fig. 5.35). At smaller depths, the slopes are somewhat uniform (0–1; 1–2 m). At bigger depths, low-value slopes are more present (Fig. 5.36). This is determined by the high clogging degree, which managed to eliminate the humps and slopes. A very important ecologic characteristic for the lacustrine basins is represented by the orientation of the banks (slope orientation) according to the cardinal points (Fig. 5.37), shown by chart (Fig. 5.38). The orientation of the banks towards the sun or in the shadow of the sun controls the distribution of the lacustrine organisms.

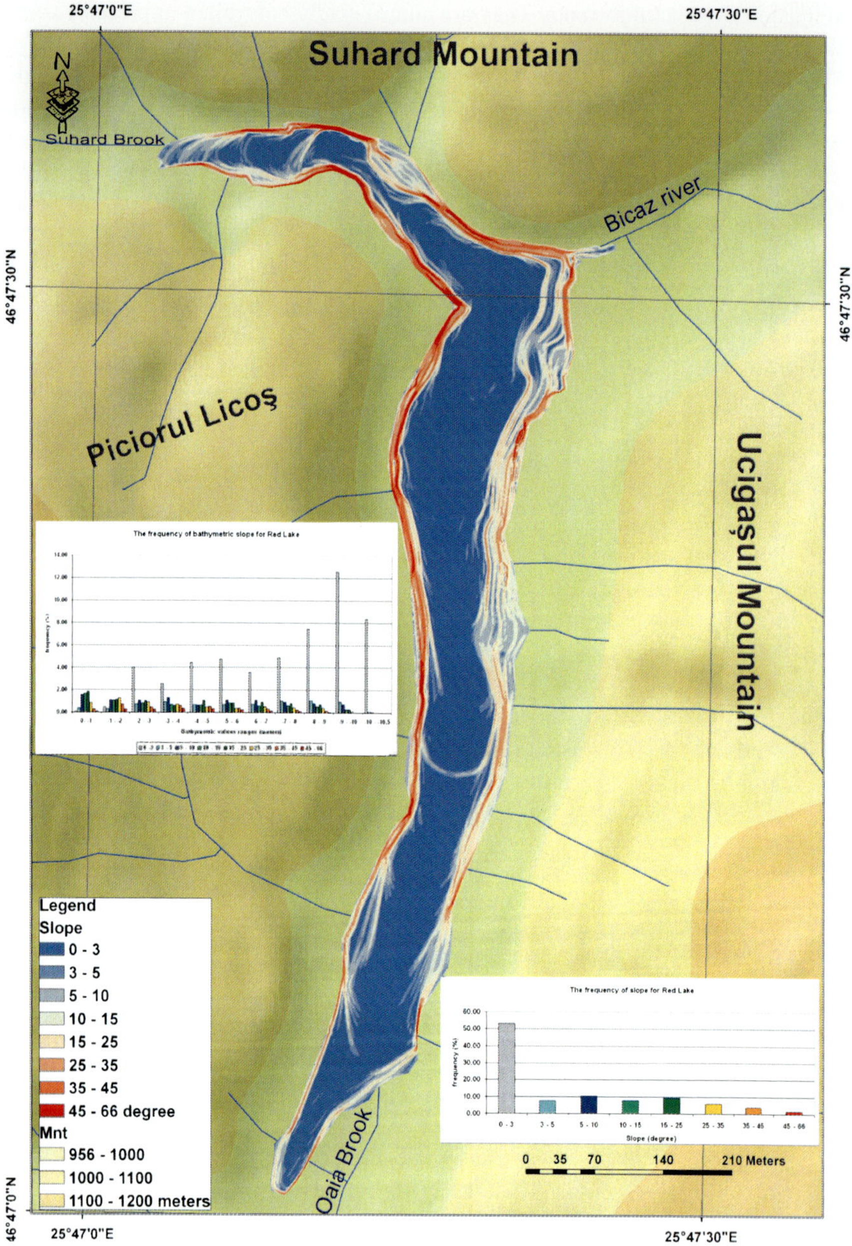

Fig. 5.34 Value of the slopes in the Red lake basin

The highest values are characteristic to the banks with eastern, northeastern and northwestern exposure (Fig. 5.38). They follow the general orientation of the lake and its elongated shape. The lowest values are characteristic to the southern and

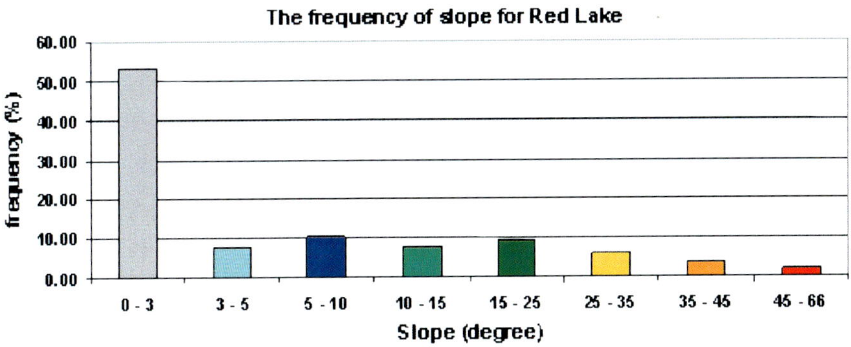

Fig. 5.35 The average distribution (%) of the slopes in the Red lake basin

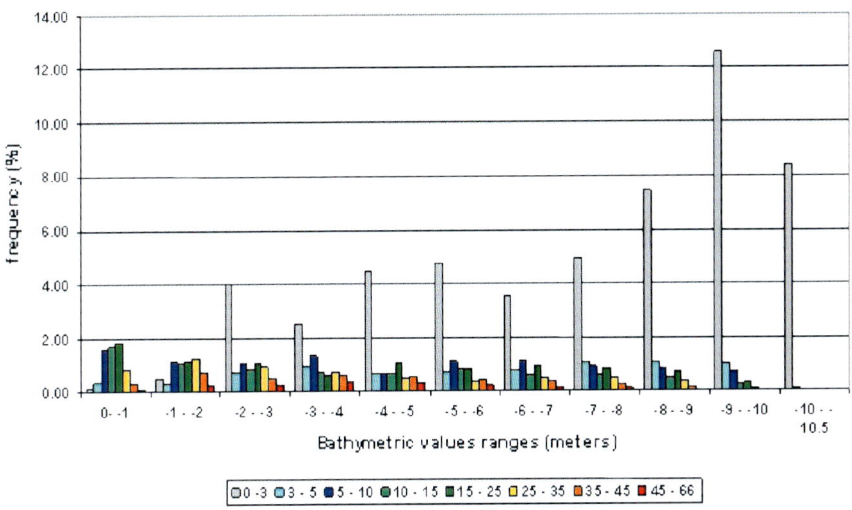

Fig. 5.36 Distribution (%) of the slopes for each depth interval

southwestern banks. Such a cartographic representation is absent from all the limno-ecological and even geomorphologic studies. The community of organisms develops on a slope or another of the lacustrine basin depending on the hydro-ecologic and morpho-bathymetric parameters. The development, along the meridians, of the biggest branch of the basin, on the south-north alignment, influences the slope orientation and the highest values are those of the slopes with an eastern and a westward orientation, respectively.

There are important differences in the exposure of banks at different depths. The general situation is somehow changed in the middle part of the lake, between the depths of 4–5, 5–6, 6–7 m, where other exposures are predominant (Fig. 5.39).

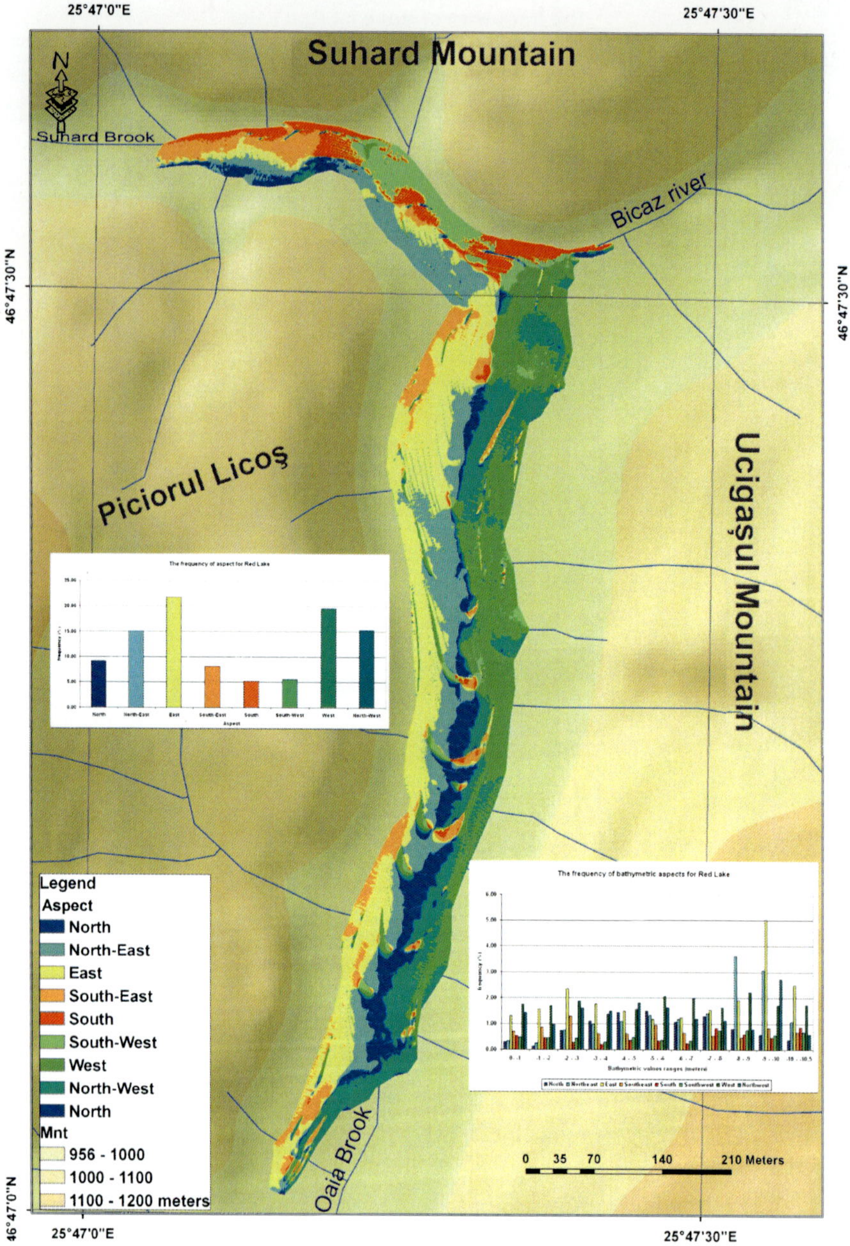

Fig. 5.37 Map of the banks orientation (slope orientation) in the Red lake lacustrine basin

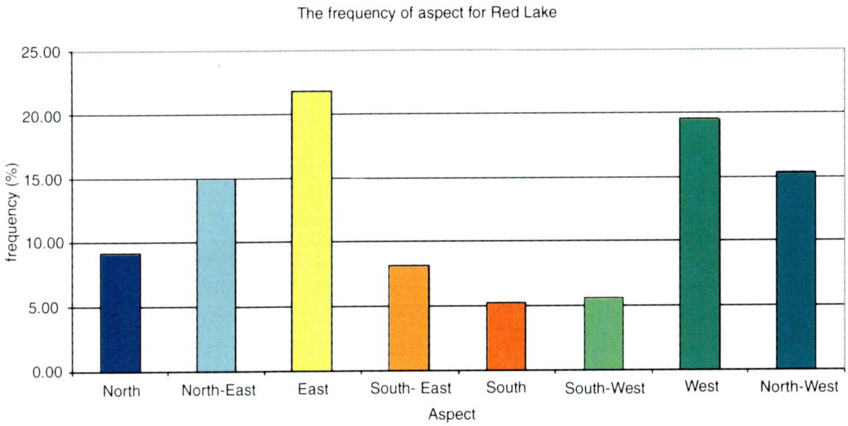

Fig. 5.38 Average distribution (%) of submerged banks (slopes) orientation in the Red lake basin

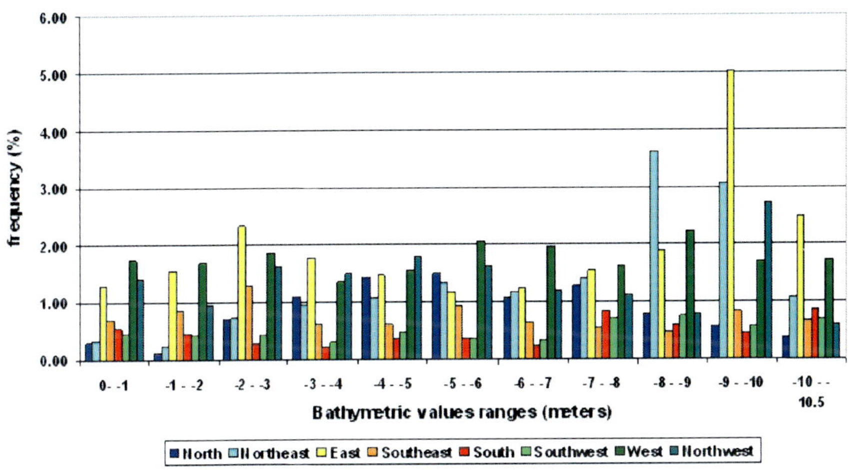

Fig. 5.39 Distribution (%) of submerged banks (slopes) orientation for each depth interval

At higher depths, the eastern and northeastern exposures are predominant. This is due to the inclination of the submerged glacis formed at the base of the rocky (hard) slope.

As regards the smaller depths, there is certain uniformity in the orientation of all slopes. At bigger depths (8–9, 9–10 m) it should be mentioned the slopes with a north-eastward, eastward and south-eastward orientation, on the left side of the basin, where the walls have high slopes. The slope affected by the landslide, with mild slopes, has a general westward orientation, while the opposite bank has an eastward general orientation (Fig. 5.40).

Fig. 5.40 The area with reduced depths on the slope affected by the landslide body

In a longitudinal profile, one can notice the relatively high bluff in the sector between the Oii stream and the river mouth (10.5 at 958.89 m). The slope of the sector between the Suhard stream and the river mouth is 10.5 at 403.32 m. In the latter case, the slope is much steeper and therefore, great depths are maintained during the whole stream. The general slope of the water level along the Oaia (Oii) stream—river mouth profile is 0.015 % and the slope of the lake bottom is 1.11 %. On the other hand, for the sector between the Suhard stream and the river mouth, the slope of the water level is 0.245 % and the slope of the lake bottom is 2.84 %.

In the case of the two confluent valleys, the slopes are different because their lengths are highly unequal (Fig. 5.41). For a greater length of the longitudinal profile belonging to the Oaia valley, the slope is smaller. For the longitudinal profile of the Suhard valley, the slope is bigger, hence the higher runoff velocity. The two streams join behind the dam formed by the landslide diluvium. The circular currents in this place lead to maximum depths and, at the same time, they create high slopes on the inside wall of the landslide diluvium.

The most developed levels are located in the first half of the Oaia (Oii) valley, where alluvia are strong. The first level, with depths even down to 2 m, is about 200 m long. The second level, with depths between 2 and 6 m, is about 300 m. The crests of the two levels are obvious; they may have been caused by the existence of hard rocks (steep slopes or crests), or of heads of landslide waves (Fig. 5.42).

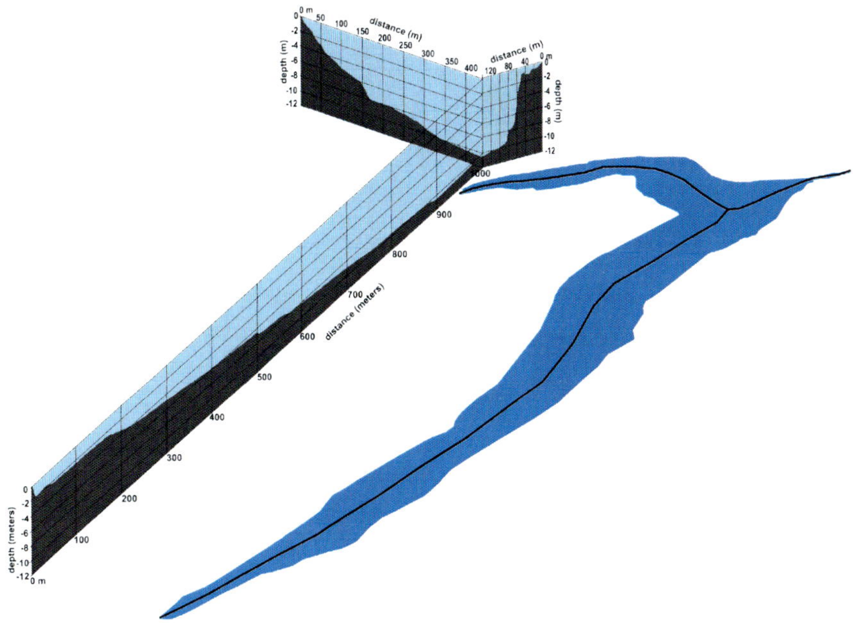

Fig. 5.41 Diagram of the longitudinal profile in the Red lake (the proportion of shapes and lengths for the two components—Oii stream and Suhard stream—is maintained)

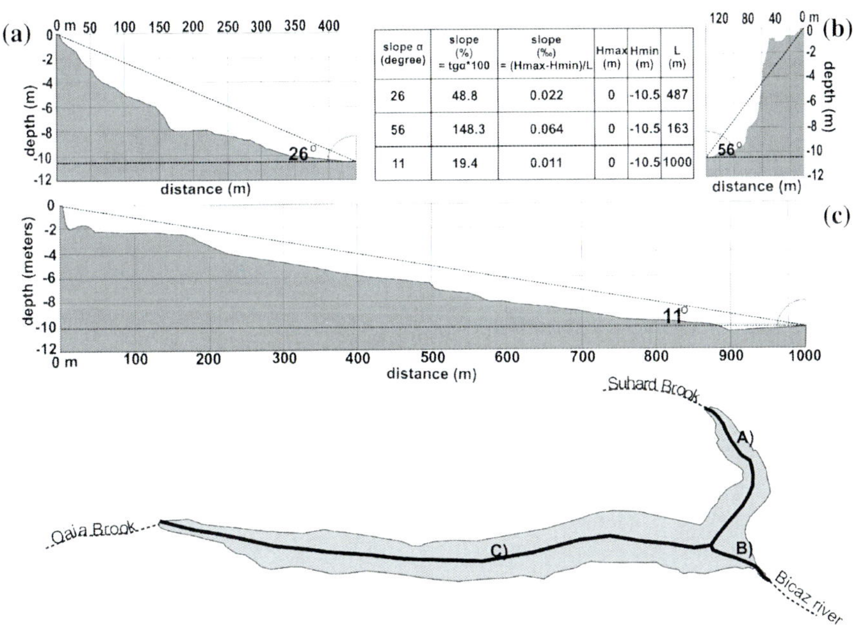

slope α (degree)	slope (%) = tgα*100	slope (‰) = (Hmax-Hmin)/L	Hmax (m)	Hmin (m)	L (m)
26	48.8	0.022	0	-10.5	487
56	148.3	0.064	0	-10.5	163
11	19.4	0.011	0	-10.5	1000

Fig. 5.42 General slope of the *bottom* of the Red lake basin

The longitudinal profile of the Suhard, of 487 m length, has an average slope of 26°. The longitudinal profile of the Oaia (Oii) valley, for 1,000 m, has an average slope of 11°. A special case is that of the slope belonging to the landslide body, forming the closing dam of the Bicaz valley. It has the shape of a wall, with an average slope of 56° (Fig. 5.42).

The landslide body impounding the Red Lake basin presents a slope with vertical values, well defined, with little change across time. Precisely these weak changes permitted the maintenance of the water volume and the existence of the lake itself.

The greatest depths, of 10.5 m, can be found at the confluence of the two inflows, Oaia (Oii) and Suhard, behind the body enclosing the lacustrine basin. The significant depths in this sector could be explained by currents of lake bottom circulation. These currents can be generated when the longitudinal currents turn back after hitting the dam. In this case, under-washing of the riverbed bottom occurs and certain enclaves with great depths are created. They can have a temporary character and they are obvious only for very high velocities of deep currents. During heavy rains, great swirling currents are developed and they represent an extra supply for the lacustrine basin. The most obvious swirling occurs at the confluence of the two currents and their impact with the landslide body of the river mouth. During the quiet moments of the tributaries, the swirling forces can be diminished and the sediment discharge can temporarily reduce the small irregularities of the lake thalweg (valley line).

The vertical distribution of the depths creates a transverse profile of the lacustrine basin in the shape of a conical bag. The area within the isobath −1 m is 11,236 ha and within −10 m only 0.056 ha. At the level of each bathymetric curve, an equivalent in solar energy is received. This varies, depending on turbidity, transparency, degree of vegetation and ice sheet cover (isolated or continuous), etc.

The general slope of the lake bottom profile is 11° along the profile from the Oaia (Oii) stream to the river mouth, 26° between the Suhard stream and the river mouth and 56° along the profile of the confluence between the two sectors and the main spillway. The greater length can also explain the lower slope in the southern sector. The higher slope of the northern sector is a result of the reduced length. The high slope next to the river mouth is a result of the presence of the extremity of the landslide body.

For each bathymetric interval (0–1; 1–2 m, etc.) there is a corresponding equivalent water volume–Vp with specific physical–chemical and biological characteristics (Fig. 5.43). Certain ecologic functions can be assessed accordingly. For each interval, there is a correspondent for submerged surfaces, of the landforms affected by the exogenous factors. If the climatic parameters change with every 100 m of altitude on the submerged landforms, the same happens within the submerged landforms but on a more diminished scale. At the latitude where Romania is located, the temperature decreases by 0.6 °C for each 100 m of altitude. In the case of the Red Lake, this parameter is 1 °C for each meter of depth. The same thing happens with the other physical–chemical parameters. Living

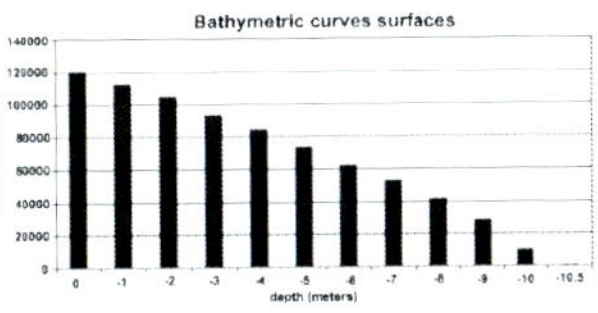

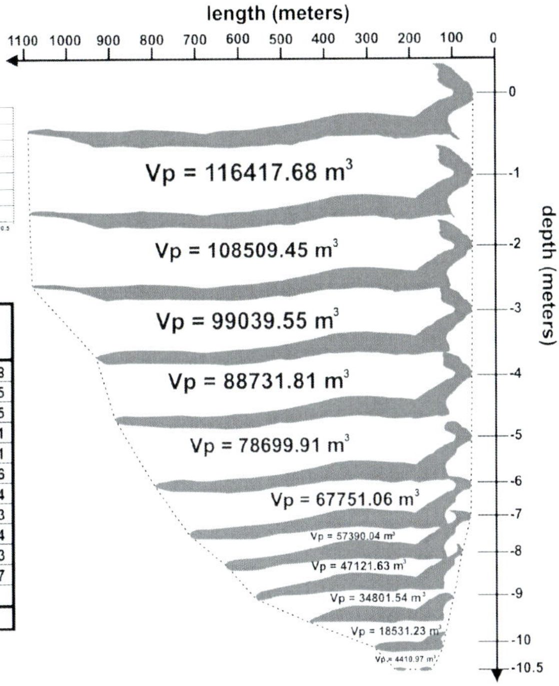

Bathymetric curve value (meters)	Surface (sq m)	Partial volume (m^3)
0	120520	116417.68
-1	112363	108509.45
-2	104701	99039.55
-3	93484	88731.81
-4	84063	78699.91
-5	73456	67751.06
-6	62202	57390.04
-7	52709	47121.63
-8	41747	34801.54
-9	28291	18531.23
-10	10263	4410.97
-10.5	564	
Total volume = 721404.87 m^3		

Fig. 5.43 Vertical distribution of *bathymetric curves* and the equivalent of the area occupied by each depth interval (1/1 m)

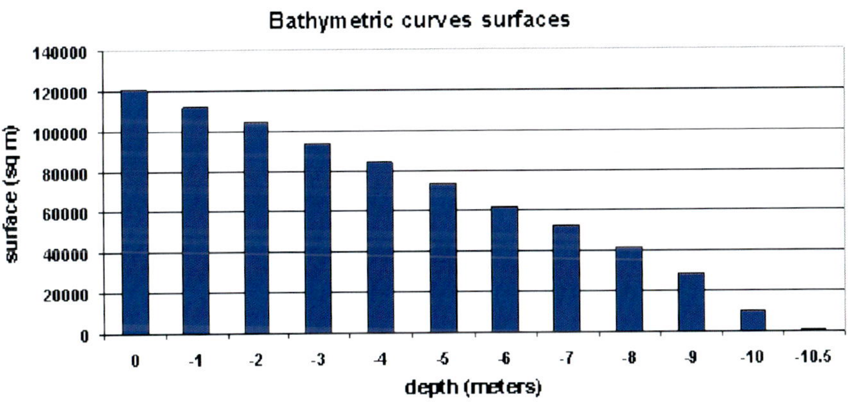

Fig. 5.44 Characteristic isobaths on the Red lake and the equivalent value of the area

conditions on the highest peaks of the mountains are severe; the same things can be said about the greatest depths of the lacustrine basins.

The total water volume of the Red Lake reaches 721,404.87 m^3. On bathymetric levels, its distribution is relatively non-uniform. The biggest proportion is that of the volume between 0 and 2 m, with a value of 224,527.13 m^3 (31.17 %). This is the sector with the most intense life. Between 0 and 4 m there is a water volume of 412,698.49 m^3 (57.20 %), where almost the entire living community of the lake can be found (Fig. 5.43). The big depth of the lake and the relatively tight valley make the water volume of big depths small. The smallest volume is occupied by the bathymetric level of over 10 m, with only 4,410.97 m^3 (0.61 %) and the biggest by the bathymetric level 0–1 m, with 116,417.68 m^3 (16.13 %).

The surface of the Red Lake is 120,520 m^2. The highest value for the surface of a bathymetric curve is a 1 m deep curve, with 112,363 m^3 (93.45 %). The lowest value is specific to the 10.5 m depth, with 564 m^2 (0.46 %) (Fig. 5.44).

In order to study the temporal evolution of the lacustrine basin, a hypsometric integral specific to the submerged landform units was made. This is based on the estimation of the degree of erosion for the landform (Fig. 5.45). The whole area of the graph represents the initial volume of the landforms. The area located under the hypsometric curve represents the present volume of landforms. The ratio between the two volumes indicates the value of the hypsometric integral (Strahler 1952). At the surface, the situation is the following: a concave curve reveals old landforms and a convex curve indicates young landforms, with weak erosion phenomena. For the Red Lake the values are higher than 1 (convex curve), which demonstrates a permanent accumulation of material and an ageing of landforms. The phenomenon is exactly the opposite of the one occurring on land.

The hypsometric integral is a graphical representation of the relationship between altitude and area (Strahler 1952). The calculation formulas for the two axes are: for the oY axis: $y = h/H$, in which $h =$ altitude of a point in a horizontal-cross section (meters) and $H =$ maximum altitude of the studied area (meters). For the oX axis: $x = a/A$, in which $a =$ the relative surface occupied (square meters), $A =$ total area (square meters).

By analyzing the hypsometric integral for the Red Lake lacustrine basin, one can notice a permanent accumulation and a continuous modification of the submerged slopes because of accretion (bottom and lateral). At the same time, the

Fig. 5.45 Hypsometric integral of the Red lake basin

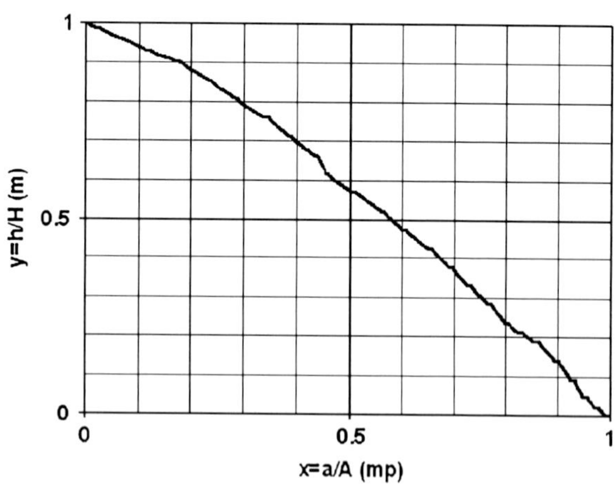

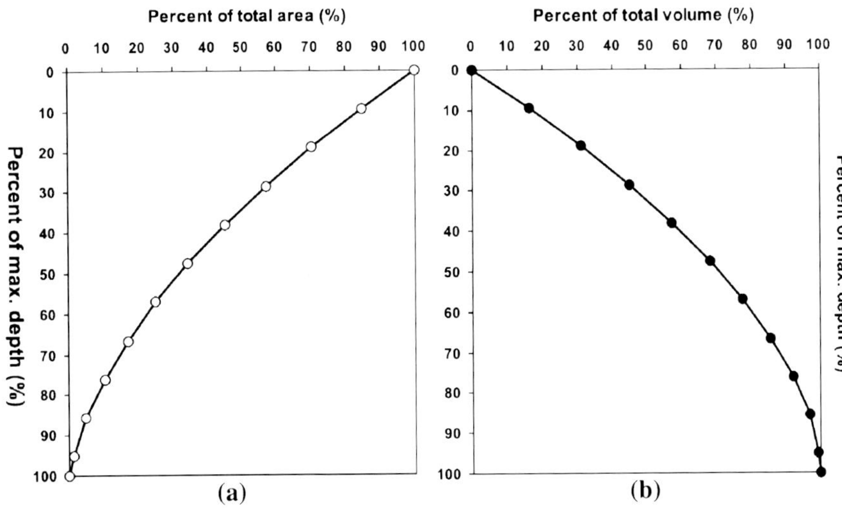

Fig. 5.46 *Hypsographic curves* illustrating the percentage of total surface area and the percentage of maximum depth (**a**), as well as the percentage of total volume and the percentage of maximum depth (**b**), of the Red lake, Harghita county, Romania

Fig. 5.47 3D model of the Red lake basin

hypsographic curves of the ratio between the area and the maximum depth (%) and the ratio between the volume and the maximum depth (%) were drawn (Fig. 5.46).

The 3D model of the lacustrine basin illustrates the correct image of the flooded valleys. The elongated shape, riverbed-like, with abrupt walls, specific to a high mountain area is underlined (Fig. 5.47). The basin has maximum development in the confluence area, where maximum depths are also present. The non-uniformities of the littoral line are given by the existence of gulfs or promontories. The gulfs correspond to the mouths of torrential bodies.

The intense sedimentation areas (the mouth of the Oaia brook) present uniform transverse profiles. In the central sector, where the depths are big and the sedimentation rate is lower, the transverse profiles are asymmetrical, being more inclined on the left.

On the S–N alignment, between the mouth of the Oaia (Oii) valley and the outlet, the longitudinal profile presents two levels (thresholds), at a distance of 612 and 812 m (Figs. 5.48 and 5.49). They correspond to tougher rocks within the

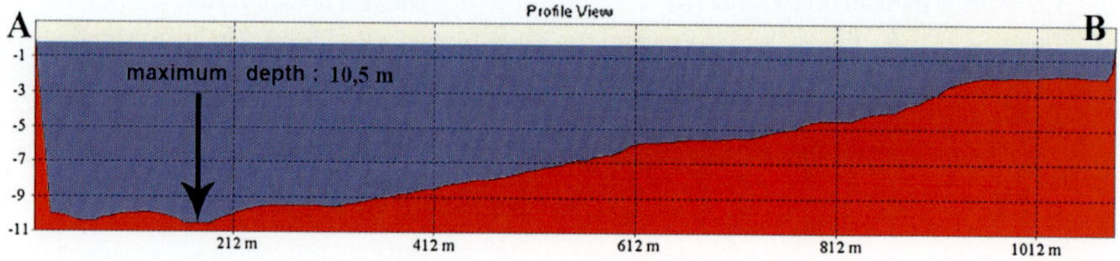

Fig. 5.48 Longitudinal profile in the Red lake basin

Fig. 5.49 The sector between the Suhard brook and the river mouth. In the background—Ghilcos Mountain and the landslide body that created a barrage in the water stream

Fig. 5.50 The mouth of the Oaia (Oii) brook and the sedimentation area

substrate. The maximum alluvium deposition in the south of the lacustrine basin created an extremely large, relatively flat surface, full of hygrophilic vegetation, such as in wetlands (Fig. 5.50).

References

Baena R, Alessia M, Fowler S, Miguel JC (2007) Particulate organic carbon: natural radionuclide ratios in zooplankton and their freshly produced fecal pellets from the NW Mediterranean (MedFlux 2005). Limnol Oceanogr 52(3):966–974

Bojoi I (1964) Observaţii asupra reliefului carstic din Masivul Hăghimaş. Sesiunea Ştiinţifică a Universităţii "Alexandru Ioan Cuza", Iaşi, (manuscript)

Bojoi I (1968a) Observaţii asupra reliefului carstic din masivul Hăghimaş. Proceedings of the Station for the Biologic, Geologic and Geographic Research Stejaru, 1

Bojoi I (1968b) Contributii la sedimentologia Lacului Rosu. In: Proceedings of the station for the biologic, geologic and geographic research Stejaru (in Romanian), pp 1, 87–105

Bojoi I (1969) Rolul proceselor periglaciare in modelarea reliefului Muntilor Haghgimas. Lucrarile Statiunii de Cercetari Stejaru, Biologie-Geologie-Geografie 2:81–110

Bojoi I (1970a) Aspecte ale evoluţiei geomorfologice a văilor din bazinul hidrografic al Bicazului. Geology-Geography Studies and Researches, Museum of Natural Sciences Piatra Neamţ, 1

Bojoi I (1970b) Procese carstogenetice actuale ale Munţilor Hăghimaş. Coroziunea depozitelor carbonatate. In: Proceedings of the Stejaru Station, Geology-Geography, Pângăraţi, 3

Bojoi I (1971a) Tipurile de doline din zona carstică a Munţilor Hăghimaş. In: Proceedings of the Stejaru Station, Geology-Geography, Pângăraţi, 4

Bojoi I (1971b) Munţii Hăghimaş şi zona Cheilor Bicazului. Studiu geomorfologic cu privire special asupra carstului. PhD thesis. Faculty of Geography and Geology, Iaşi

Bojoi I (1971c) Particularités du Karst des Monts Hăghimaş (Carpates Orientales). Revue Studia Geomorphologica Karpato-Balkanica, Krakow

Bojoi I (1972) Rolul fenomenelor carstice în evoluţia Cheilor Bicazului. Institutul Pedagogical Oradea, Geography Section, Oradea

Bojoi I (1976) Precizări privind influenţa litologiei şi structurii asupra reliefului Munţilor Hăghimaş. Annals of the Museum of Natural Sciences Piatra Neamţ, Geology-Geography series, 3

Bojoi I (1987) Succesiunea temporală şi concomitentă în modelarea prin carstificare-crioplanaţie a platourilor calcaroase ale Masivului Hăghimaş. Scientific Annals of the Alexandru Ioan Cuza University Iaşi, sII, Geology-Geography, 33

Bojoi I (1988) Formarea depresiunilor carstice în legătură cu etapele de evoluţie paleogeografică a Masivului Hăghimaş. Scientific Annals of the Alexandru Ioan Cuza University Iaşi, sII, Geology-Geography, 34

Bojoi I (2000) România. Geografie fizică. Alexandru Ioan Cuza University Publishing House, Iaşi

Catana C (1989) Studiu mineralogic şi petrografic al Unităţii Gnaiselor de Rarău-Hăghimaş. Teză de doctorat, Alexandru Ioan Cuza University, Iaşi

Cazacu C, Brustur T, Szobotka S, Melinte MC (2009) Studii asupra sistemelor hidrografice in aria Parcului National Cheile Bicazului—Hasmas. GEOECOMAR Session of Scientific Communications, Constanta (in Romanian), pp 171–180

Ciaglic V (2005) Contributii la cunoasterea caracteristicilor hidrologice si hidrogeologice ale muntilor Ghilcos-Haghimas-Mezinul si a conditiilor care le determina. Terra Nostra Publishing House, Iasi (in Romanian), pp 1–71

Clague J, Luckman BH, Van Dorp RD, Gilbert R, Froese D, Jensen BJL, Reyes AV (2006) Rapid changes in the level of Klaune Lake in Yukon Territory over the last millennium. Quaternary Research, Elsevier 66:342–355

Dragoş V (1957) Deplasări de teren. Scientific Publishing House, Bucharest

Gagnon P, Scheibling RE, Jones W, Tully D (2008) The role of digital bathymetry in mapping shallow marine vegetation from hyperspectral image data. Int J Remote Sens 29(3):879–904

Gâştescu P (1971) Lacurile din Romania—Limnologie regionala. Romanian Academy Publishing House, Bucharest (in Romanian), pp 1–372

Gâştescu P (1979) Lacurile Terrei. Albatros Publishing, Bucharest

Gâştescu P (1998) Hidrologie. Roza vânturilor Publishing House, Târgovişte

Giurgiu M (1990) Peşterile din Masivul Hăşmaş. Buletin CSER, Baraolt

Godet L, Fournier J, Toupoint N, Olivier F (2009) Mapping and monitoring intertidal benthic habitats: a review of techniques and a proposal for a new visual methodology for the Europa coasts. Prog Phys Geogr 33(3):378–402

Grasu C, Turculeţ I, Catana C, Niţă M (1995) Petrografia Mezozoicului din "Sinclinalul Marginal ". Romanian Academy Publishing House, Bucharest

Grasu C, Miclăuş C, Brânzilă M, Baciu DS (2010) Munţii Hăşmaşului. Monografie geologic şi fizico-geografică. Al.I.Cuza University Publishing House, Iaşi

Grasu C, Miclăuş C, Brânzilă M, Baciu DS (2012) Sinclinalul Mezozoic Tulgheş-Hăşmaş-Ciuc. "Al.I. Cuza" University Publishing House Iaşi

Horton RE (1945) Erosional development of streams and their drainage basins: hydrophysical approach to quantitative morphology. Geol Soc Am Bull 56(3):275–370

Ichim I (1973) Cu privire la unele fenomene periglaciare din Carpaţii Orientali. Realizări în Geografia României, Scientific Publishing House, Bucharest

Ichim I (1976) Aspects de la cryoplanation et de la cryopedimentation des Carpathes Orientales Roumaines. Revue Roumaine de Geologie-Geophysique-Geographie, serie Geographie, Bucharest 22

Islam MS, Hashim R (2010) Behavior of stabilized peat: A field study. Sci Res Essays 5(17):2366–2374

Mayer R (1932) Bericht uber morphologische Studien in den Ostkarpathen. Anuarul Institutului de Geologie al Romaniei 18:56–68

Mfundisi KB (2008) Soil fertility in wetland versus reclaimed land using plant parameters in relation to nitrogen content: the case of Yala Swamp Western Kenya. Sci Res Essays 3(7):287–293

Mujdeci M, Kara B, Isildar AA (2010) The effects of different soil tillage methods on soil water dynamic. Sci Res Essays 5(2):3345–3350

Naum T, Preda I (1972a) Morfolitologia şi morfostructura zonei masivului Hăghimaş. Bulletin of the Romanian Society of Geographic Sciences 2(LXXII)

Naum T, Preda I (1972b) Relieful calcaros şi relieful carstic din Munţii Hăghimaş. Lucrările Ştiinţifice ale Institutului Pedagogical Oradea, Geography series, Oradea

Nechita N (2003) Flora si vegetatia cormofitelor din masivul Hasmas, Cheile Bicazului si Lacului Rosu (The flora and vegetation of the cormophytes from Hasmas Massif, Bicaz Gorges, and the Red Lake). Bibliotheca Historiae Naturalis II, Natural Sciences Museum, Piatra Neamt, pp 1–383

Orghidan T, Puşcariu V, Bleahu M, Decu V, Rusu T, Bunescu A (1965) Harta regiunilor carstice din România. Lucrările Institutului Speologic Emil Racoviţă, Bucharest, 4

Pandi G (2004) A Gyilkos-Tó. Hidrogeográfiai tanulmány. Casa Cartii de Stiinta Publishing House, Cluj, pp 1–231

Pişotă I, Năstase A (1957) Lacul Rosu, nod de confluenta a trei bazine hidrografice. Probleme de geografie (in Romanian) 4:181–205

Popp N (1941) Cronica Geografica, A) Fapte. Congresul Profesorilor de Geografie de la Piatra Neamt. 18–21 May 1940, Bulletin of the Romanian Society of Geography, Bucharest, LX, 317–342

Preda I, Naum T (1973) Contribuţii la cunoaşterea apelor minerale din Munţii Hăghimaş. Scientific Proceedings of the Pedagogical Institute Oradea, Geography Series, Oradea

Puşcarciuc R (1988) Consideraţii preliminare asupra endocarstului Masivului Hăşmaş. Montana bulletin, Oneşti

Rădoane M, Rădoane N (1986) Consideraţii asupra reliefului format pe gresii şi conglomerate din România. Annals of the Museum of Natural Sciences Piatra Neamţ, Geology-Geography, Piatra Neamţ 5

Romanescu G, Cojocaru I (2010) Hydrogeological considerations on the western sector of the Danube Delta–a case study for the Caraorman and Saraturile fluvial-marine levees (with similarities for the Letea levee). Environ Eng Manag J 9(6):795–806

Romanescu G, Dinu C, Radu A, Torok L (2010a) Ecologic characterization of the fluviatile limans in the south-west Dobrudja and their economic implications (Romania). Carpathian J Earth Environ Sci 5(2):25–38

Romanescu G, Stoleriu C, Lupascu A (2010b) Morphology of the lake basin and the nature of sediments in the area of Red Lake (Romania). Ann Univ Oradea: Geogr Ser 20(1):44–57

Săndulescu M 1975 Studiul geologic al părţii central şi nordice a sinclinalului Hăghimaş (Carpaţii Orientali). Yearbook of the Romanian Geologic Institute, Bucharest, XLV

Strahler AN (1952) Hypsometric (area-altitude) analysis of erosional topography. Bulletin Geological Society of America 63:1117–1142

Udriste O (1963) Lacul Rosu si imprejurimile. Meridiane Publishing House, Bucharest (in Romanian), pp 1–55

Walkley A, Black IA (1934) An examination of the Degtjareff method for determining organic carbon in soils: effect of variations in digestion conditions and of inorganic soil constituents. Soil Sci 63:251–263

Zăvoianu I (1978) Morfometria bazinelor hidrografice. Romanian Academy Publishing House, Bucharest

Chapter 6
The Nature of the Sediments Within the Lacustrine Basin

Abstract The sediment discharge that ends up in the Red Lake is important, mostly given that the last decades have been characterized by intense deforestations. These have accelerated the erosion and the silting of the lake basin, as well. In order to correctly assess the source areas of the sediments, from the drainage basin, satellite imagery has been used, to accurately identify the areas that have suffered from deforestation in the last decades. Through deforestation, the slopes are severely destabilized, being vulnerable to landslides, high-water waves, etc. Several silt samples from the bottom of the lake were taken in 17 points covering the whole surface of the lake. The thickness of the lacustrine sediments varies extremely, reaching a maximum of over 6 m in the spillway sectors of the two important brooks: Oaia (Oii) and Suhard. The silting process was also measured by comparing bathymetrical transects between the measurements taken by Pişotă and Năstase in 1957 and the measurements taken in 2010 with specialized equipment. Therefore, the silting rate is a very important parameter, associated with numerous processes both natural and anthropic. Silting has also been studied with the help of gravity core samples. Several USLE models have been made, to emphasize the difference between natural erosion and accelerated erosion, due to illegal deforestation. Silting analysis is very important, because, by using different methods, predictions on future silting rates can be made and we can estimate the lifespan of the lake.

The number of streams alimenting the Red Lake waters is relatively high. In this case, the solid discharge (in suspension and crawled) is important, mostly given that the last decades have been characterized by intense deforestations. The friable rocks are susceptible to erosion and the clogging rate is relatively high.

The samplings were used for the analysis of the organic carbon and the humus—elements necessary for the hygrophilic and hydrophilic plants within the lake (Fig. 6.1). In the total organic carbon determination, to obtain the percent of carbon content from the soil, first the titrant solution ($FeSO_4 \cdot 7H_2O$) must be standardized before the sample analyses are made. The resulting data were reduced and the following equation was used:

G. Romanescu et al., *Limnology of the Red Lake, Romania*,
DOI: 10.1007/978-94-007-6757-7_6,

Fig. 6.1 Samplings of silt from the lake bottom

Equation (6.1), Titrant normality equation:

$$N_2 = \frac{N_1 V_1}{V_2} \tag{6.1}$$

In which:

N_1: $K_2Cr_2O_7$ normality,
V_1: $K_2Cr_2O_7$ volume (mL),
V_2: $FeSO_4$ volume (mL).

Equation (6.2), Organic carbon percentage:

$$\%\,Carbon = \frac{(A - B) \times 0.3 \times 1.33}{C} \tag{6.2}$$

where:

A: meq $K_2Cr_2O_7$ = (mL $K_2Cr_2O_7$ × N $K_2Cr_2O_7$),
B: meq $FeSO_4 \cdot 7H_2O$ = (mL $FeSO_4 \cdot 7H_2O$ × N $FeSO_4 \cdot 7H_2O$),
C: grams of sample,
0.3: conversion factor to carbon weight.

The result of the difference between A and B is in milli-equivalents and they need to be converted to carbon milli-equivalents in order to obtain the units needed. For this, it is necessary to perform the next operation:

Equation (6.3),

$$\text{meq} \times (1\text{eq}/1000\text{meq}) \times ((1/4)\text{molC}) \times (12\text{g}/1\,\text{molC}) \times 100 = 0.3\,\text{gC}(\%) \tag{6.3}$$

The 0.3 conversion factor has units of carbon grams and involves the constant to convert a fraction to percent units; hence, Eq. (6.2) does not have the factor 100. The Walkey-Black constant for the sediment of 75 % is the mean recuperation of carbon in solids and sediments by using this method. That is why the final results have to be multiplied by 1.33 in order to get the real value, this constant is not used when determining carbon in KHP standard because almost all its carbon content is recovered.

In order to transform the organic carbon into humus, it needs to multiply ii by the factor 1.724. In 1837, because of continuous summer precipitation, an immense landslide occurred on the western slope of Ucigasu Mountain, creating a barrier in the stream of the Bicaz brook. The landslide wave favored the appearance of a natural barrage lake, which extended on the streams of the Oaia (Oii) and Suhard valleys. It has a total area of 12.1 ha (120,134.44 m^3) and a water volume of 721,404.87 m^3.

At the water level, in a longitudinal profile, one can notice a more reduced slope in the sector of the Oaia (Oii) valley, from 965.199 m upstream, to 965.051 m downstream, at the river mouth and a higher slope on the Suhard valley, from 966.041 m upstream, to 965.051 m downstream, at the river mouth. The situation is similar at the level of the lake bottom, with the difference that, at the river mouth, the altitude is 954.55 m, 10.5 m lower—which is an equivalent of the lake maximum depth. The general slope of the lake bottom profile has the value of 11° on the alignment of the Oaia brook and the river mouth, 26° between the Suhard brook and the river mouth and 56° on the alignment of the confluence between the two sectors and the main spillway. Soil erosion is influenced by the way the land is used and the extent to which the hydrographic basin is covered by vegetation, implicitly (Fig. 6.2). Clogging is also influenced by the (mostly illegal) deforestation of relatively large areas, especially pine spruce areas. In this case, certain aspects should be briefly presented relating to the evolution of deforestation.

The method used makes reference to the photo-interpretation of satellite images. The satellite images were downloaded from the Global Land Cover Facility site (http://glcf.umiacs.umed.edu/); they share satellite images covering the whole globe. Satellites capture the satellite images with special sensors, capable of separating precise channels of the electromagnetic spectrum. By using the download option of this internet address, first it localizes the area of interest and then the satellite images for several years are downloaded. The first data available go back to 1989 and five sets of satellite images were downloaded (1989, 2001, 2002, 2005 and 2007). The satellite images are available for several periods of the year but, unfortunately, not all years are covered. This is the reason why a non-interpolation analysis was made. The deforestations do not represent a natural process with predictable cycles but a process determined by the anthropic factor,

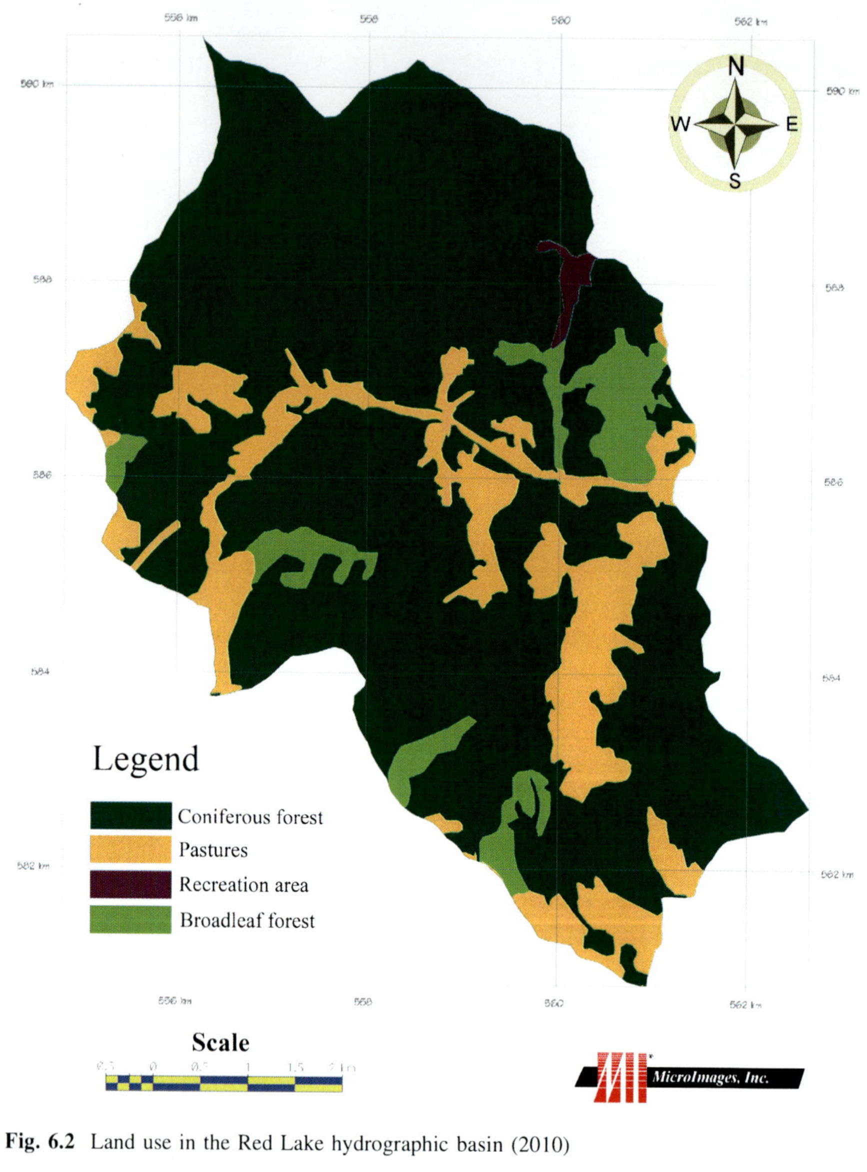

Fig. 6.2 Land use in the Red Lake hydrographic basin (2010)

with an unpredictable temporal variability. In order to obtain the most impressive results, the sets of satellite images downloaded are dated exclusively from the summer months (July, August), when there is a lot of vegetation, in order to delimit, in a highly accurate manner, the deforested areas from one year to another.

These areas were emphasized by vectoring the deforested areas (Fig. 6.3), depending on the color composites formed from separate channels within the RGB additive color system and from infrared variations (thermal, closed) within the

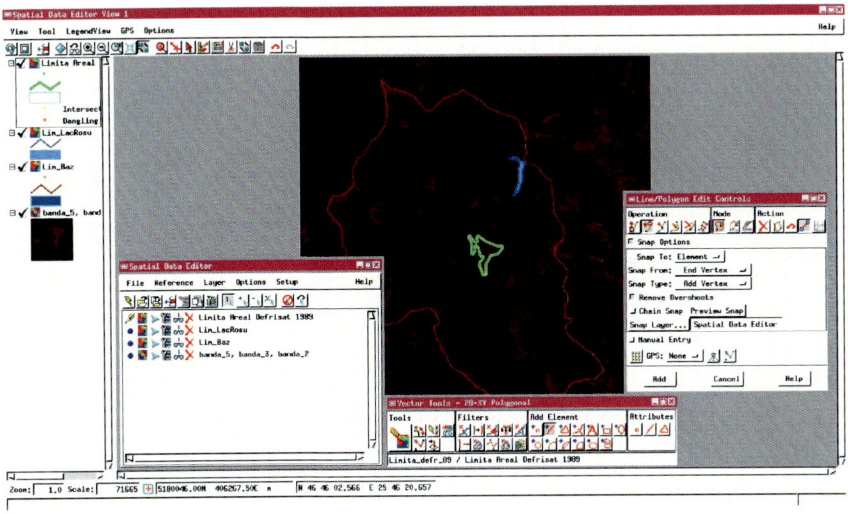

Fig. 6.3 The working method in TNT Maps for digitizing the deforested areas

electromagnetic spectrum. These very last (infrared) channels have favored the creation of certain composites with an obvious delimitation between the areas with shrubby vegetation and those with a predominant mineral material (nude rock).

Though the fidelity of the graphic information is not optimal, the materials obtained may be used in order to demonstrate the hypothesis and to understand the bigger picture concerning the deforestation activities within the catchment basin of the Red Lake. In order to synthesize the information at a chrono-spatial level, for a better understanding, a graph—underlying the differences between the deforested surfaces within consecutive annual data sets—was generated.

There is an oscillatory trend of the deforestation activities. The measurement intervals have the following values: 1989–2001 = 2.2 km^2; 2001–2002 = −0.72 km^2; 2002–2005 = 2.18 km^2; 2005–2007 = 1.26 km^2 (Fig. 6.4). There was an increase in the deforestation rhythm between 2001 and 2002.

The annual means of the deforested surfaces are 18 km^2/year for the interval between 1989 and 2001; 0.72 km^2/year for 2001–2002 (with more intense deforestations); 0.72 km^2/year for 2002–2005; and 0.63 km^2/year for 2005–2007 (Fig. 6.5). The calculation may be considered reliable, as it has not taken into account the replanted (even-aged) forests, which present another chromatic nuance on the satellite images. Chances are very low that, within the annual intervals, trees were cut and then replanted several times. Even for the first interval, covering 16 years, the development of trees could have been observed if new (even-aged) forests had appeared. There is no evidence whether the deforestation is concentrated in a short period or if it has a uniform temporal distribution. As regards the interval between 1989 and 2001, it is not certain that the deforestation activities were concentrated in only 2–3 years.

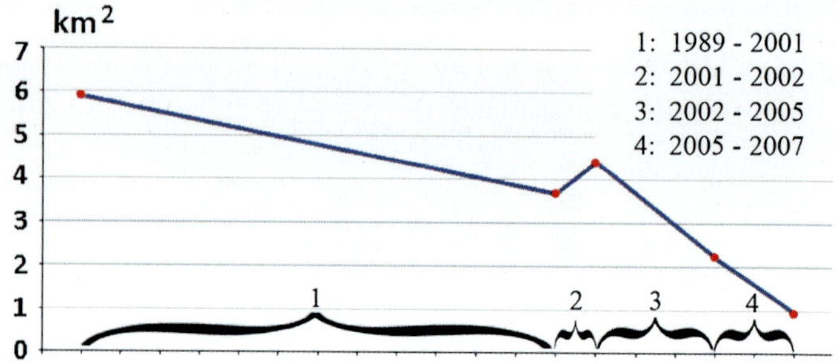

Fig. 6.4 Evolution of the deforested surfaces during 1989–2007

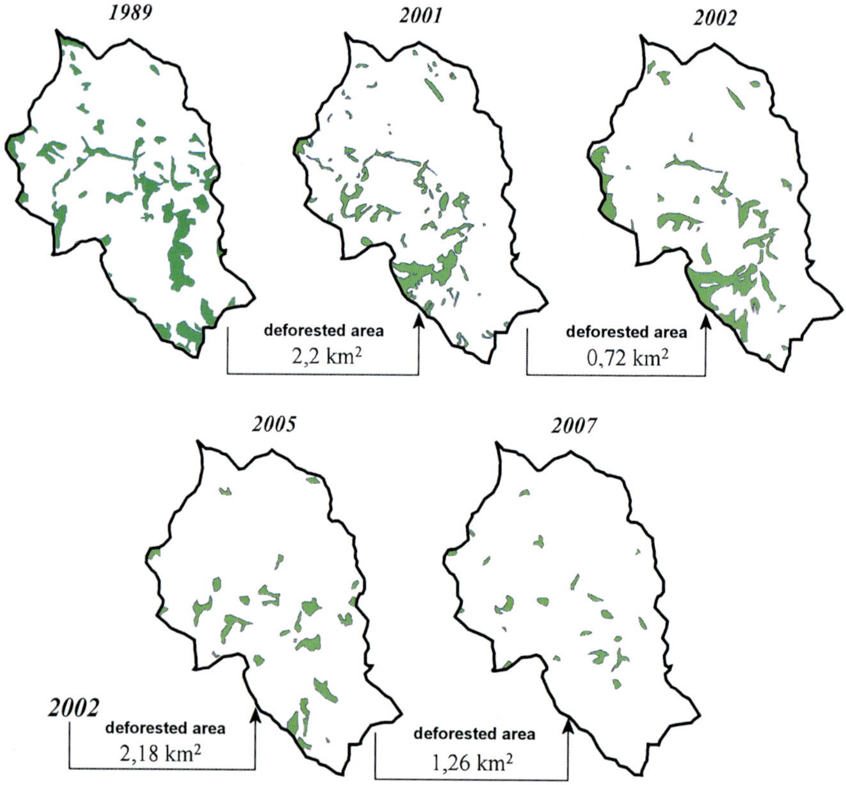

Fig. 6.5 Localization of the deforested surfaces for the interval between 1989 and 2007

Most deforested surfaces are located in the southern half of the basin. In the northern half, the forests were conserved due to the Red Lake—tourist attraction of national interest. In this case, the deforestations have been strictly controlled as, landscape-wise, they would not contribute to promoting local tourism. In the northern half, there is a national road (12C), which helps to maintain a low-deforestation level. Illegal deforestation is easier in the hidden, non-circulated areas. In the southern half, there are only forest roads, with strict local importance. Between 1989 and 2007, the deforestations extended over the watershed in the southern part of the basin, which proves that this phenomenon has a much larger dispersion (Fig. 6.6).

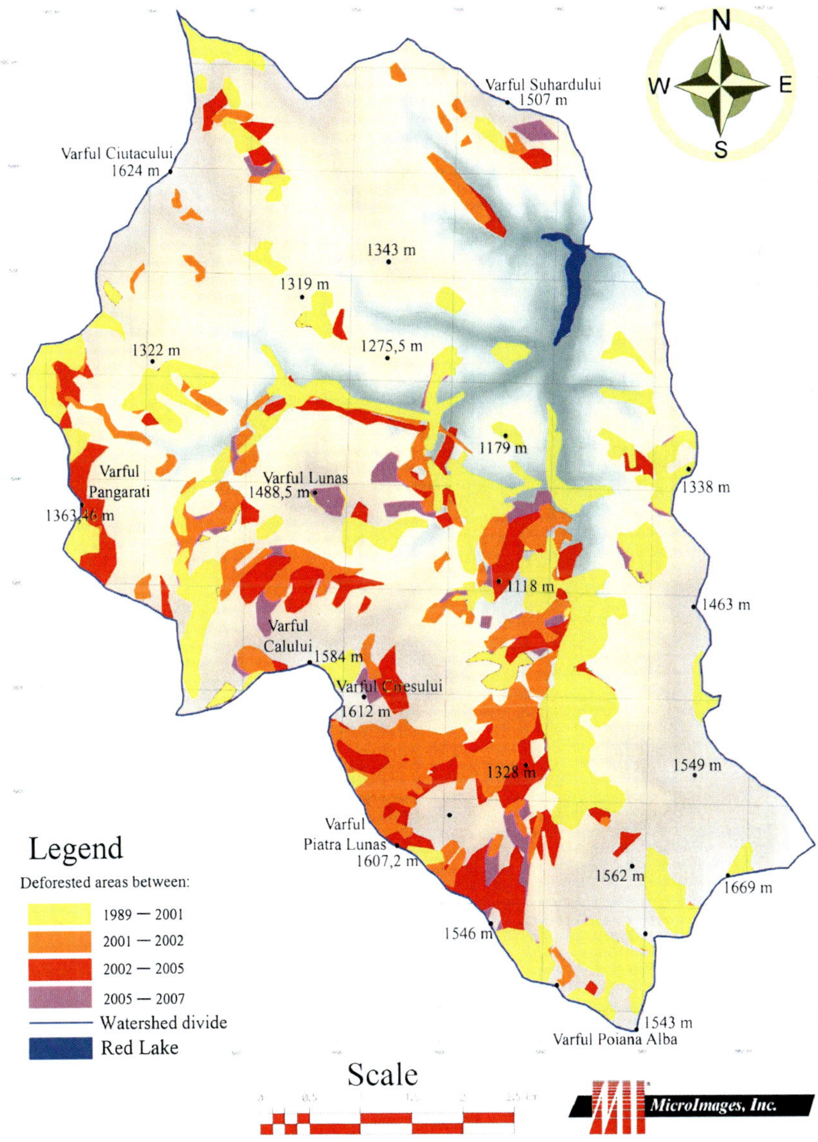

Fig. 6.6 The deforested surfaces between 1989 and 2007

Some of the deforestations within the center of the basin are oriented along the road, probably because of the maintenance works executed in the area or because there was a need for more visibility for traffic. Most of the deforested surfaces within the southern sector are situated on the lower side of the slopes on the valley of the Oaia (Oii) brook. In this case, there is a qualitative and quantitative accentuation of the pluvial denudation process, given the lack of a vegetal layer that could reduce the erosive action of rainwater and could prevent the gravitation displacement of dislocated soil particles, fixing them in situ. The empty slopes are easily washed. The alluvia reaching the Oaia (Oii) brook are transported by it to the Red Lake.

Through deforestation, the slopes are severely destabilized, being vulnerable to landslides, high water waves, etc. The silt samples from the bottom of the lake were taken in 17 points covering the whole surface of the lake (Fig. 6.7). These points are representative of the erosive processes at the surface, as well as for the alluvial processes within the lake basin.

As water flowing speed reduces, the brooks tributary to the Red Lake gradually deposits alluvia: first, the rolled and dragged ones and then the suspension ones. The former are deposited mainly next to the source and the latter are deposited further away. The lacustrine sediments in the area of the Oaia (Oii) brook spillway have a grey color, with a grey-blackish color towards the surface. The darker color at the surface is a result of higher humus content. Next to the Suhard brook an abundance of craggy elements is registered. In this case, the slope of the river is greater as well. The sediments in the Red Lake present the characteristics of fine micaceous sand.

The thickness of the lacustrine sediments varies extremely, reaching a maximum of over 6 m in the spillway sectors of the two important brooks: Oaia (Oii) and Suhard. The quantitative variations of the organic carbon and the humus are extremely large, mostly in the complicated structures of the Alpine–Carpathian–Himalayan chain (Miller et al. 2009; Mulholland and Kuenzler 1979).

One can notice a logical correlation between the existence of the total organic carbon and humus quantities (Table 6.1, Fig. 6.8).

The more reduced the slopes and depths, the higher the concentration of organic carbon and humus is (sample points 1, 2, 3, 4, 5 and 15, 16, 17 respectively). In the deeper sectors, especially towards the river mouth, smaller quantities are found as well (Fig. 6.9). This fact is due to the great distance to the alluvia source and to the existence of a weak aquatic current that transports a reduced amount of alluvia. The alluvia are stopped at the two extremities of the lake by the existence of immerse and submerged vegetation, extremely well developed in the torrential rain season.

The greatest value of organic carbon is 7.31 % at the spillway of the Oaia (Oii) brook and 6.10 % at the spillway of the Suhard brook. The most reduced value (3.73) is registered next to the river mouth, where the greatest depths are registered as well. In the field literature, values of over 25 % are cited at the river mouth of the Oaia (Oii) brook and 18 % at the spillway of the Suhard brook (Bojoi 1968a, b). In the present case, it seems that these values are exaggerated.

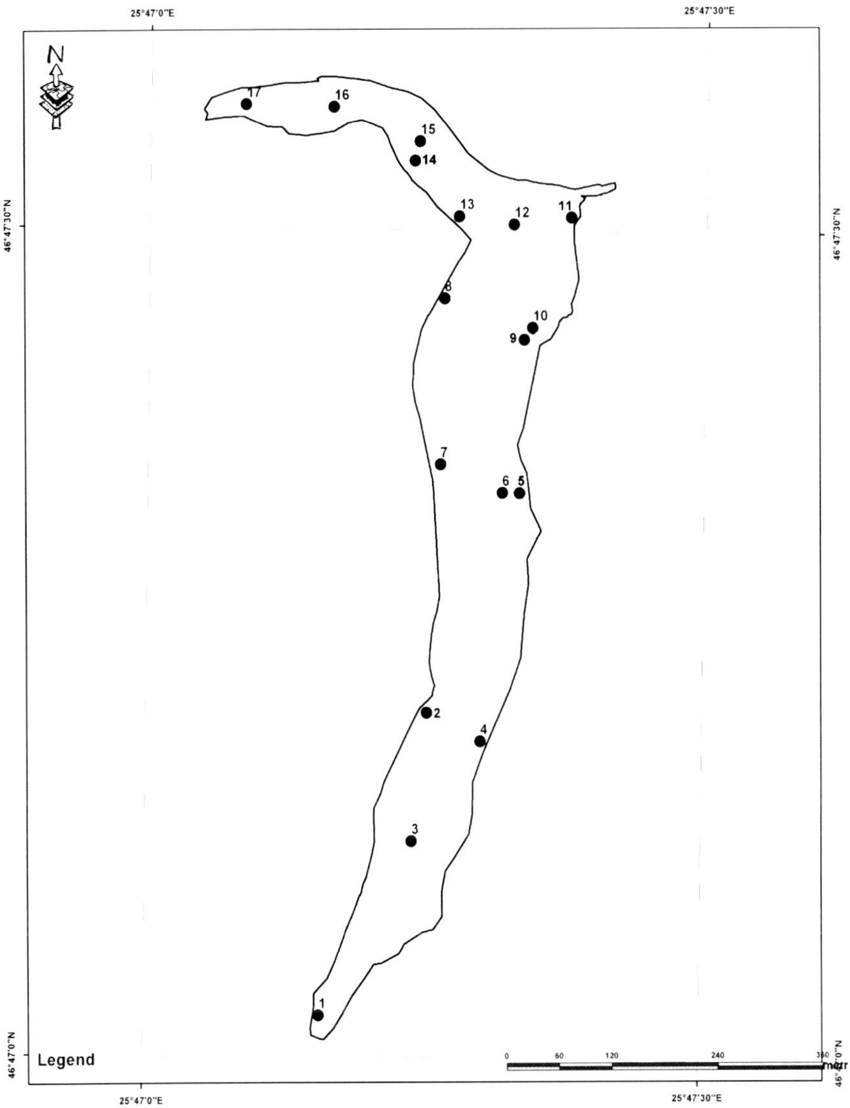

Fig. 6.7 Distribution of the points where samples of silt were taken from the bottom of the Red Lake

The high values of humus are caused by the existence of rich vegetation in the sectors with reduced depths. In these areas, the water mass is totally lighted and the roots of the plants can penetrate the soft layer. These are the most important wetlands around the lake (Fig. 6.10).

The humus in the sediments of the lake comes from two sources: eroded soil from the slopes; accumulation produced by plant decomposition. This way a

Table 6.1 Amount of total organic carbon and humus in the sediments of the Red Lake

Sample	Total organic carbon %	Humus %
1	7.31	12.7
2	6.61	11.39
3	4.03	6.94
4	3.82	6.58
5	6.65	11.46
6	4.49	7.74
7	4.26	7.34
8	4.21	7.25
9	3.92	6.75
10	3.73	6.43
11	4.80	8.27
12	3.82	6.58
13	4.78	8.24
14	4.21	7.25
15	5.06	8.72
16	5.39	9.29
17	6.10	10.53

typical limnic soil is formed. During summer, when torrential rains occur, the brooks transport an important amount of alluvia material. The yellow color, which indicates the high amount of alluvia, is specific only to the sectors next to the riverside spillway. The rest of the lake remains clear, with extremely weak turbidity. The transparency degree reduces to several centimeters next to the banks and in the area of abundant vegetation (10–20 cm), while the sectors close to the river mouth and the sectors with the greatest depths have a transparency of 2–3 m.

The greatest values of humus are 12.7 at the spillway of the Oaia (Oii) brook and 10.53 at the spillway of the Suhard brook. The lowest value is 6.43 and it is registered at the river mouth, where the greatest values of depth are also registered. The existence of a significant amount of humus and typical limnic soil, implicitly, facilitated the development of immerse and submerged vegetation (Fig. 6.11). This fact is normal for depths up to 2–3 m. For greater depths, light hardly penetrates and therefore, the bottom of the lake is lacking such life forms. This is the domain of the bacteria and animals that can resist under such conditions created by oxygenation, pH, salinity, pressure, etc.

On the lithological slopes (western and northern) where the slopes are greater and the depths are also greater, the lowest quantities of organic carbon and humus are found. This fact is due to the weak erosion of the slopes (perfect forest cover) and to the lack of brooks (therefore a lack of supplementary alluvia). The deposition of the alluvia material coming from the slopes and from the organism decomposition, contributes to the silting of the lacustrine basin and to the changing of the whole morphology of the Red Lake, implicitly (Fig. 6.12).

The sedimentary deposit in the Red Lake basin represents a formation specific to mountain valley lakes. An important characteristic of these alluvia is

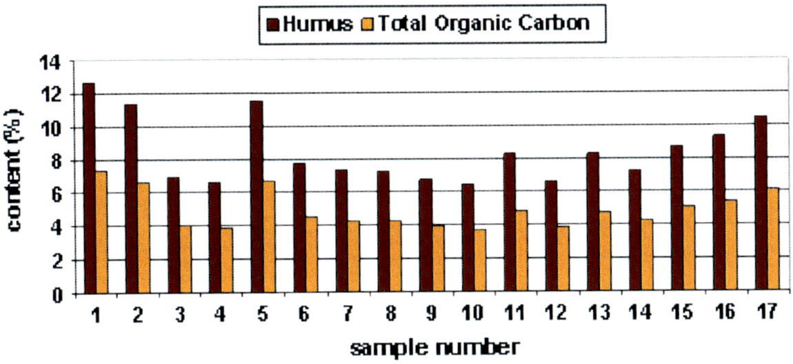

Fig. 6.8 Histogram of the distribution of total organic carbon and humus proportions in the sediments of the Red Lake

represented by the existence of sands with intercalations of coarse fragments, specific to the changes in the sedimentation process. Most of the sediments come from the geologic substrate made of crystalline schists—on the western flank of the syncline—sandstones, calcareous sandstone Doggers and wildflysch.

The amount of organic carbon and humus implicitly increases towards the surface. This can be explained by the development of vegetation while the depth lowers. In order to study the temporal evolution of the lacustrine basin, a hypsometric integral specific to the immerse landform units was made. This is based on the estimation of the landforms erosion degree. The whole area of the graph represents the initial volume of the landforms. The area situated under the hypsometric curve represents the present volume of landforms. The ratio between the two volumes indicates the value of the hypsometric integral (Strahler 1952).

At the surface, the situation is the following: a concave curve reveals old landforms and a convex curve indicates young landforms, with weak erosion phenomena. For the Red Lake the values are higher than 1 (convex curve), which demonstrate a permanent accumulation of material and an aging of landforms. The phenomenon is exactly the opposite of the one occurring on land.

The hypsometric integral is a graphical representation of the relationship between altitude and area. The calculation formulas for the two axes are:

For oY axis: $y = h/H$, in which $h =$ altitude of a point in a horizontal-cross section (meters) and $H =$ maximum altitude of the studied area (meters). For oX axis: $x = a/A$, in which $a =$ the relative surface occupied (square meters), $A =$ total area (square meters).

By analyzing the hypsometric integral for the Red Lake lacustrine basin, one can notice a permanent accumulation and a continuous modification of the submerged slopes because of accretion (bottom and lateral). Human induced changes around the lake—as important sources of sediments—are extremely rare. It should be mentioned there are only several deforested areas on the southwestern slope and an area with several tourist constructions on the north-eastern sector.

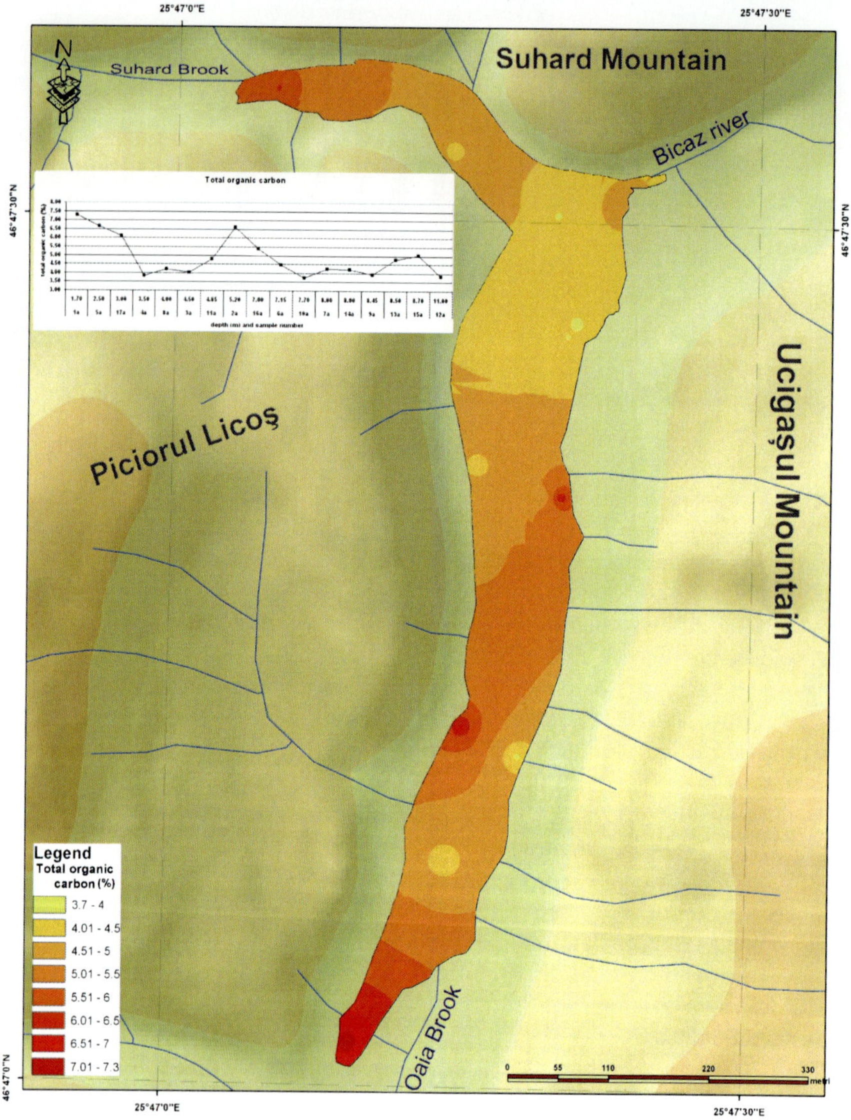

Fig. 6.9 Map of total organic carbon distribution in the surface sediments of the Red Lake

All the lacustrine basins represent the main depository of local erosion. It is clear that the erosion rate is different from one area to another, depending on the complexity of the physical-geographic factors (Clague et al. 2006; Forrest et al. 2008).

Unfortunately, there are no subsequent studies of the sedimentation rate. Even the new studies have a high degree of generality (Begy et al. 2009). The only data are those noted by Bojoi in 1968. The average progradation rate, at the mouth of

Fig. 6.10 Wetland corresponding to the Oaia (Oii) brook

the Oaia (Oii) brook was 2–5 m/year. The sediments are 5–6 m thick. The average clogging rhythm at the level of the Oaia (Oii) is 2.8–8.2 cm/year. Macrophyte vegetation plays an important role in sedimentation (Nechita 2003).

At the level in 1968, the volume of the sediments accumulated in the lacustrine basin rose to 480,000 m^3 (Bojoi 1968a, b). Granulometrically, the following fractions can be distinguished: sand—from coarse sand to fine sand, 87.5 %; silt—6.5 %; gravels—6 %. The clogging degree has reached 40 % (Bojoi 1968a, b). There are important bathymetric differences between 1957 and 2010. The most obvious changes are noticeable at the outlet of the brooks: Oaia (Oii) and Suhard (Fig. 6.13). The lowest clogging index has been registered at the Red Lake outlet (0.9 m in the interval between 1957 and 2010), where the two hydrographic arteries join: Oaia (Oii) and Suhard. The circular currents active in this sector make the depths today identical with those of 1957 (Fig. 6.14).

The most intense aggradations occur at the outlet of the Oaia (Oii) brook, with average rates of 1.5–1.6 m, in the interval between 1957 and 2010. The high sedimentation rate is due to the existence of friable rocks and to the massive deforestations within the southern part of the hydrographic basin. This way, the Oaia (Oii) brook is loaded with an important quantity of solid material, which it unloads at the lake entrance (Figs. 6.15, and 6.16). Besides the main brooks, there are numerous torrential bodies carrying each year important quantities of alluvia, deposited, most of the times, in shapes such as fan-delta (Fig. 6.17).

The maximum thickness of the sedimentary within the Red Lake basin may exceed 6 m. The variations are obvious at the mouths of the main tributaries and in

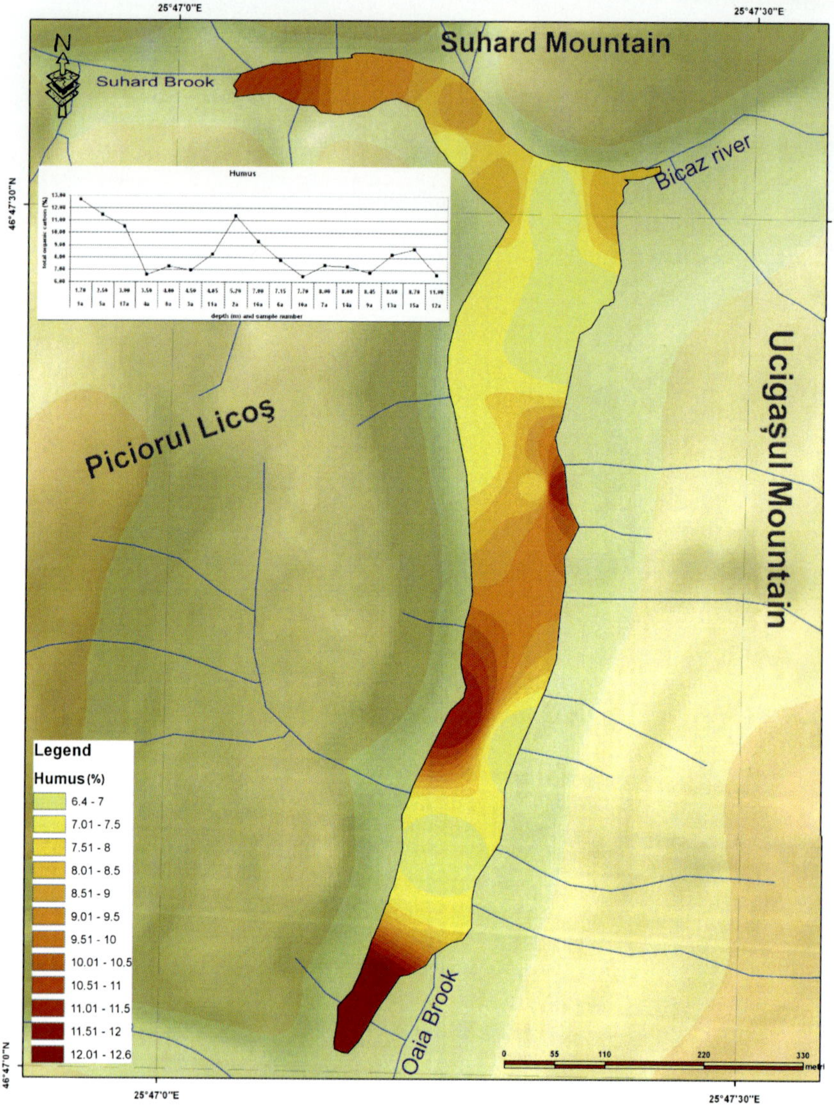

Fig. 6.11 Map of humus distribution in the surface sediments of the Red Lake

the central part of the lacustrine basin. The distance to the outlets and to the central sectors may also influence the granulometric distribution of the alluvial deposit. It appears that the sedimentary material occupies over 40 % of the initial volume of the lacustrine basin (Bojoi 1969).

The sedimentation was more intense until the catchment dams on tributaries were built (in the 1960s). Afterwards, there was an obvious decrease in the average sedimentation rate. From the granulometric perspective, the predominant

Fig. 6.12 Wetland corresponding to the Suhard brook

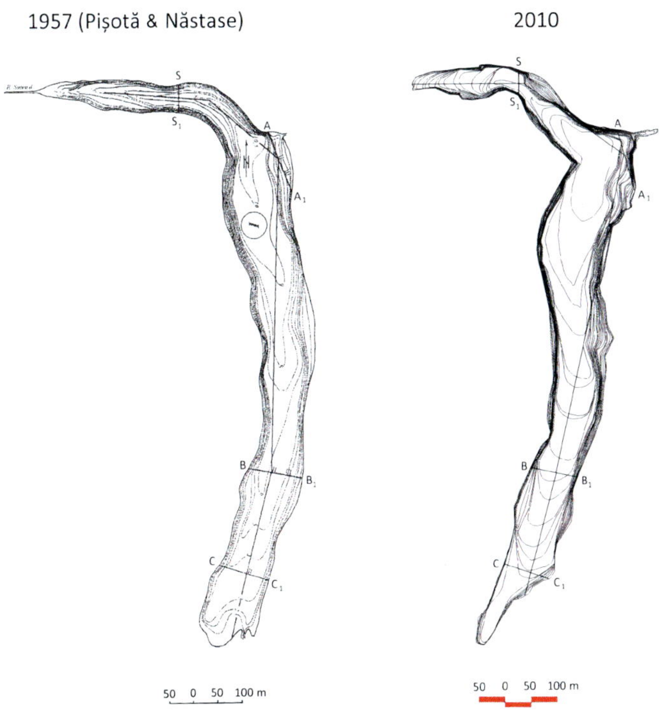

Fig. 6.13 Bathymetric comparisons between 1957 and 2010

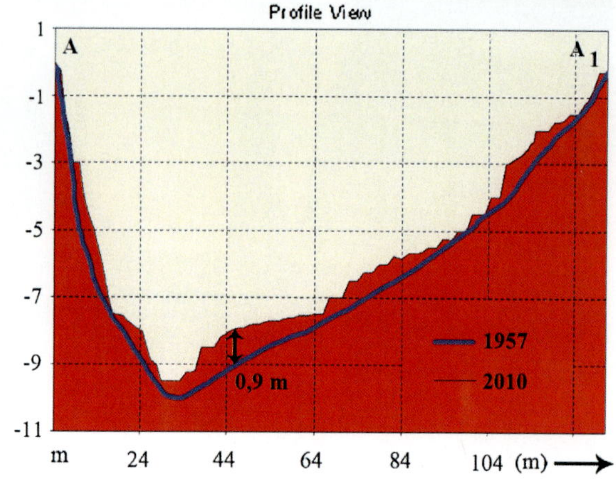

Fig. 6.14 Accumulation rate in the downstream sector, at the Red Lake outlet

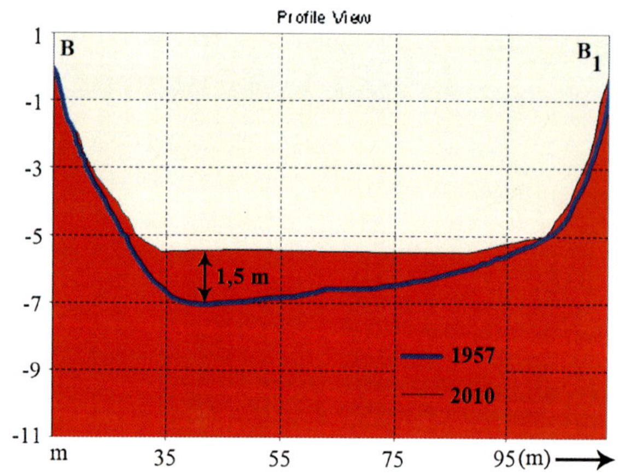

Fig. 6.15 Average sedimentation rate in the upstream sector, at the outlet of the Oaia (Oii) brook

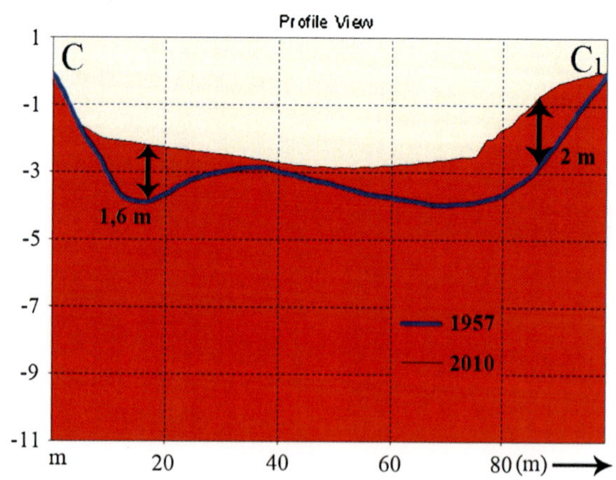

Fig. 6.16 Average sedimentation rate at the outlet of the Oaia (Oii) brook

Fig. 6.17 Alluvial fan in the south-western sector of the lacustrine basin

sediments belong to the average and fine classes of sand (Bojoi 1968a, b). The shape of the granulometric curves demonstrates the existence of frequent climatic and hydro-geomorphologic changes. The alluvia accumulated in the Red Lake basin are included into a rather large variety of granulometric classes: gravel (6 %), coarse sand (30 %), average sand (24 %), fine sand (33.5 %) and siltstones (6.5 %).

The coarse sedimentary materials are specific to the deposit in the base and the finer materials—to the upper side. The thickest gravel deposits belong to the outlet of the Suhard brook, which also has a longitudinal profile with a bigger slope. The finer materials are found in the downstream sector of the Oaia (Oii) brook. In this case, the phenomenon is obvious in the current sedimentation, after building the catchment dams for the alluvia, in the downstream sector of the brooks coming from the south.

For a correct analysis of the deforestation outcomes, a series of USLE (Universal Soil Loss Equation) models were generated, specific for four distinct types of pedologic coefficients. This way, a comparative perspective of the current situation, associated with a reduced increase in the soil being eroded and of the situation with a maximum erosive potential was elaborated. The peak values of erosion oscillate between 0.83 tons/ha/year and 1.14 tons/ha/year. These values apply only for the deforested areas. These areas have been calculated by subtracting the regions with no forested surfaces in 2007, from the equivalent stratum, corresponding to the year 1989. The resulted stratum was used to create a vector that included all the information related to surfaces and erosion. The initial extension (18 August 1989) of the areas with no forested surfaces was calculated at 5.52 km^2, while on 27 July 2007, the value doubled (11.04 km^2).

The Added Erosion Rate (AER) is applied to deforested areas. The result is cumulated with the general USLE model, for each of the four pedologic coefficients. In this case, there is a wide variety of values for the Added Erosion Rate, with a maximum 114 tons/ha/year (Fig. 6.18). The extreme values are usually isolated and they do not influence the general erosion rate (5.28 tons/ha/year).

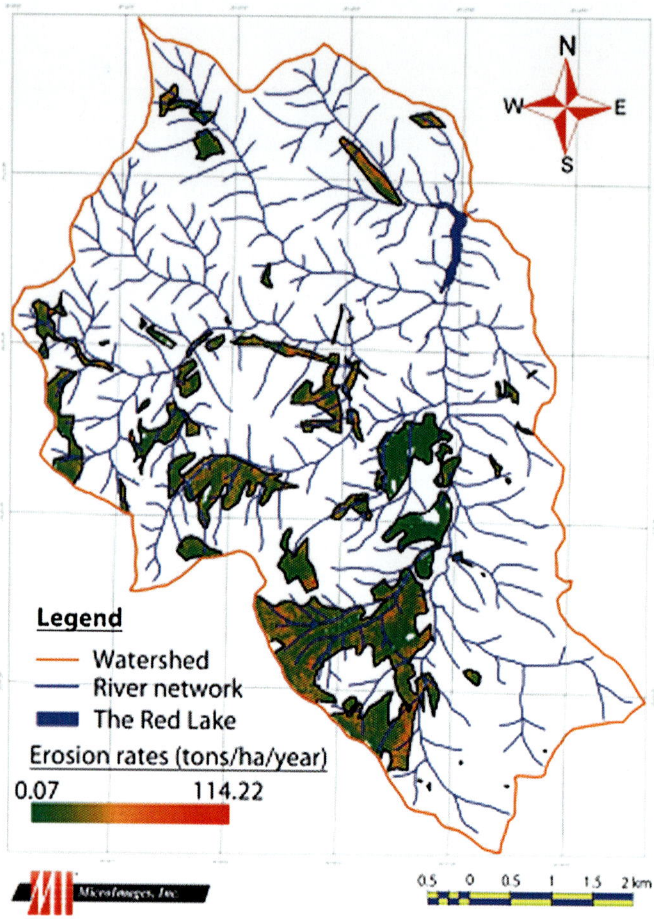

Fig. 6.18 Added erosion rate for deforested areas (2007), for a soil coefficient of 1.2

The general trend of localizing the deforested areas mainly concerns the vicinity of the principal hydrographic arteries within the drainage basin of the Red Lake. This occurs because the cut-down trees are carried by towage along the streams. It is a frequent practice because it implies less effort and money. Though illegal, it still occurs in almost all the isolated areas on the Romanian territory. The erosion rate is yet another argument for the existence of massive deforestations in the isolated areas of the catchment basin (the upstream sector, where there is low accessibility). In this case, the solution is to include the entire drainage basin within the Bicaz Gorges Hăghimaş National Park.

The drainage basin of the Red Lake has a total surface of 40.59 km^2 (4,059 ha), of which the deforested areas comprise 571 ha (14 % of the total surface corresponding to the drainage basin). For the calculations for 2007, references were made only to the erosion difference (Fig. 3). In this case, there was no change in

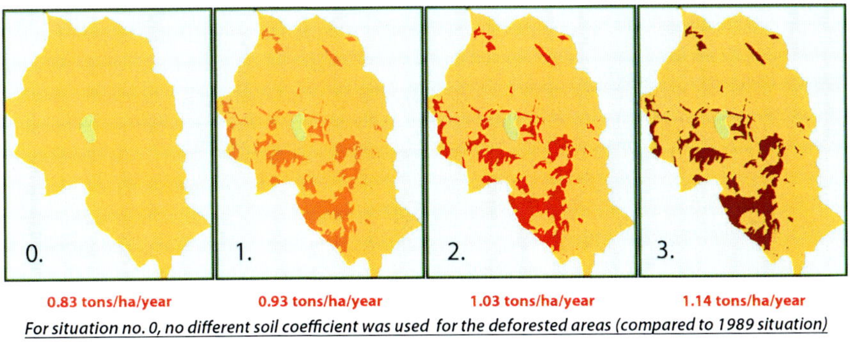

Fig. 6.19 Comparative view of the added erosion in 2007, for different soil coefficients

the pedologic coefficient (the 0.9 value—corresponding to the situation 0—remained unchanged). The changes in the use of fields (from forested surfaces to pastures) have been underlined. The result was an increase in the erosion rate by 40 % of the total erosion rate (Fig. 6.19).

The four situations for which the differences between the pedologic coefficients were determined did not reveal any major alterations. In this case, the added sediment mass represents approximately 2.5–3 % of the entire eroded sediment mass. The average erosion rate increased from 2.78 (calculated for 1989) to 4.6/5.28 tons/ha/year. This change demonstrates that 14 % of the entire surface of the drainage basin determined almost doubled the values of the erosion rate. This process took 18 years and the period corresponds to the massive deforestations during Romania's transition period (post-1989).

Because of the high clogging rate, local authorities came up with construction plans for alluvia retention reservoirs. Since 1960, three such reservoirs have been built on the lower streams of Oaia, Piatra Roşie and Licaş brooks.

Unfortunately, these basins did not prove a good long-term solution. This occurs because they have a reduced capacity to storage the sediments, due to under-sizing. The retention-clogging rate on the Oaia brook went up to 8.2 cm/year. In this moment, the retention is completely clogged. The Piatra Roşie (Vereschiu) sedimentation basin is 90 % clogged.

The indicators calculated for the four different situations, associated with the values of pedologic coefficients are underlined (Table 6.2). The general deviation (between the extreme situations) between the minimum and maximum of the coefficients is only 0.5 %, even considering the significant soil degradation.

The most significant difference in the erosion processes appears after modifying field usage. This way, the well-distributed erosion rate from the spatial perspective before 1989 was concentrated in deforested areas (Fig. 6.20). The most affected zones are localized in the downstream sector of the Oaia brook. While the pedologic stratum influenced the USLE model by just 2–3 %, the vegetation index influenced total erosion by approximately 40 %.

Table 6.2 Erosion indicators applied to different soil coefficient values (for the 2007 forest distribution)

Year	SC*	MER*	TSM* (tons)	AER*	ASM* (tons)	PI* (%)
1989	0.90	2.78	11,284.02	–	–	–
2007—Situation 0	0.90	4.60	18,671.40	0.83	472.08	2.53
2007—Situation 1	1.00	4.82	19,564.38	0.93	530.99	2.71
2007—Situation 2	1.10	5.05	20,497.95	1.03	589.89	2.88
2007—Situation 3	1.20	5.28	21,431.52	1.14	648.80	3.03

* *SC* Soil Coefficient; *MER* Mean Erosion Rate; *TSM* Total Sediment Mass; *AER* Added Erosion Rate; *ASM* Added Sediment Mass; *PI* Percentage Influence

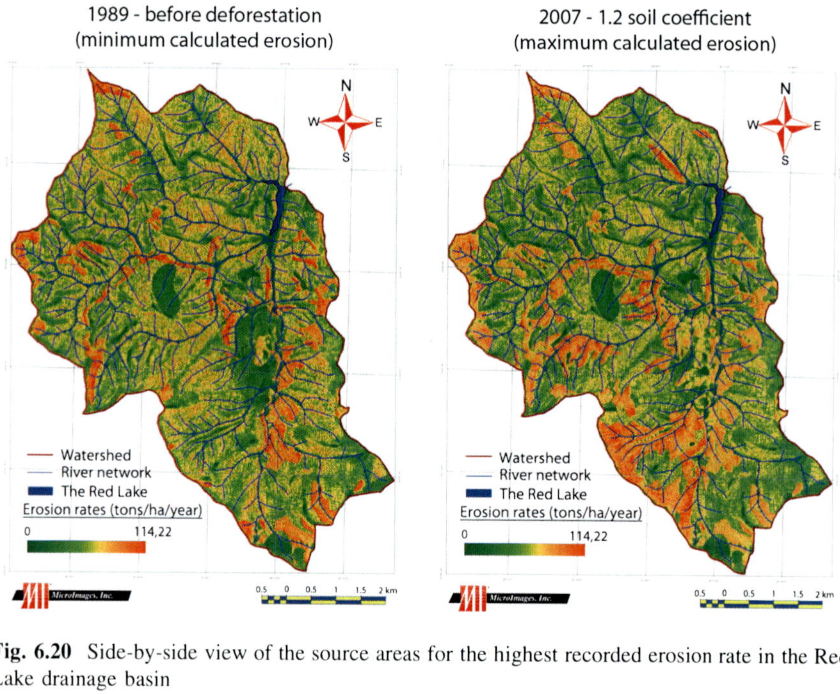

Fig. 6.20 Side-by-side view of the source areas for the highest recorded erosion rate in the Red Lake drainage basin

Another negative influence on the current situation of the Red Lake is the increase is the anthropic impact. The tourist attractions in the area—considering the beauty of the lake and of the surrounding areas—determined the construction of several cottages and pensions. The calculations reveal an alarming conclusion, which is that the erosion rate is increasing (natural erosion, rain aggressiveness, deforestations, etc.) and that it determines a massive sediment deposition in the basin of the Red Lake. Unfortunately, human intervention turned out to be the opposite of viable.

For the most part, natural dam lakes disappear rapidly because of dam break or clogging. Despite the above-mentioned tendency, the Red Lake has proved rather

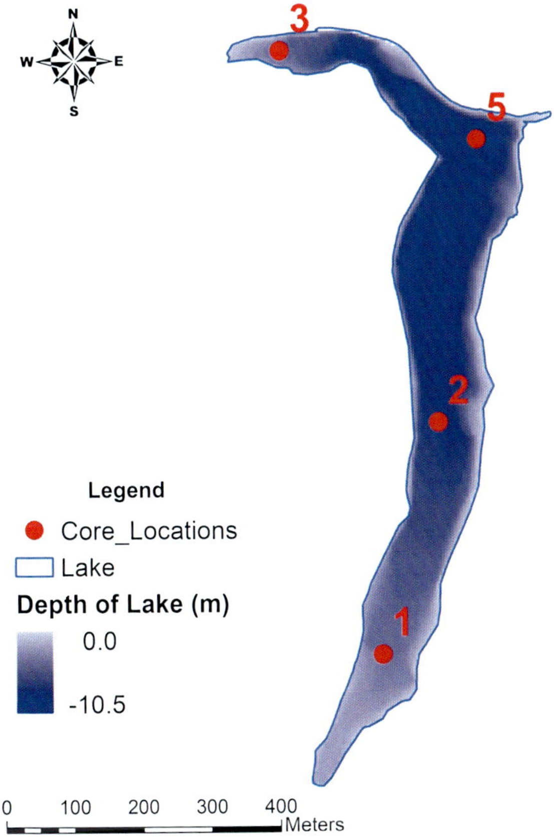

Fig. 6.21 Gravity core sample points. Cores *1*, *2* and *5* lie in the main body of the lake (running south to north). Core *3* is located in the Suhard brook arm to the west

long lasting. The main factors that have contributed to extending its life span are the following: the existence of an erosion-resistant landslide body; the reduced surface of the drainage basin (it does not facilitate the existence of a reduced amount of material); the high forestation rate; the relatively reduced amounts of precipitations; the low liquid discharge of brooks, etc.

The average erosion rate, generated through the USLE model for 2007, was 4.6–5.2 tons/ha/year. In 1989, the average erosion rate was 2.78 tons/ha/year. The ever-increasing deforested areas—besides the hydrotechnical solutions, applied to sediment retention but proven ineffective—underline that the Red Lake is in a real danger of extinction because of clogging.

Recent palaeolimnology of the Red Lake; assessing sediment sources for environmental management

The Red Lake is an iconic water body and a popular visitor destination; consequently, it has multiple user groups with an interest in its effective environmental management. The significance of forestry activities in the catchment and their impact on the siltation of the lake is also a key-question (e.g., Begy et al. 2009).

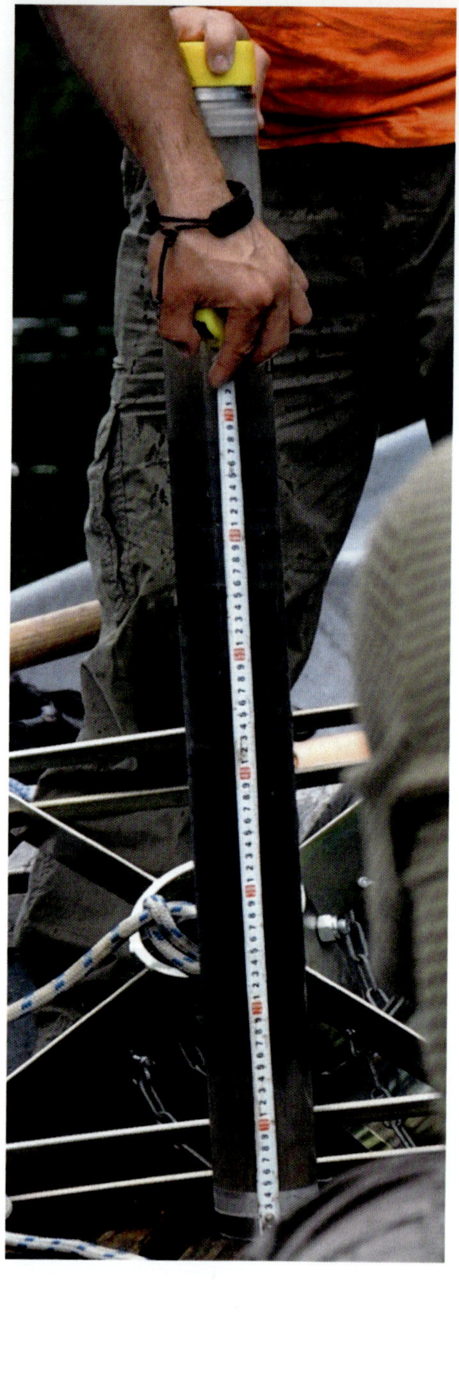

Fig. 6.22 Core 5 showing the change in sediment properties at *c*.40 cm and banding in the *upper part* of core 7

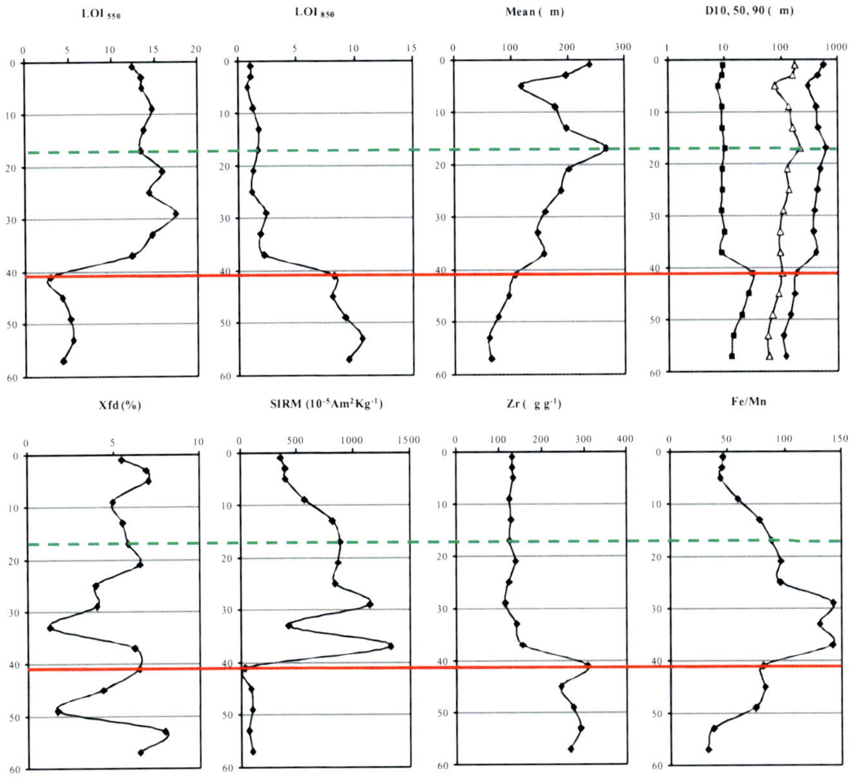

Fig. 6.23 Properties of core 5. The *solid red line* marks a major sediment boundary at 40 cm. The *dashed green line* indicates a later shift (especially in particle size) in the mid 1980s and where Cs-137 levels peak as a result of Chernobyl 'fallout'

Whereas the origins and hydrology of the lake have been well documented (e.g., Pandi 2004; Romanescu and Cojocaru 2010), the lake's recent sediment has received much less attention. This preliminary assessment aims to identify both temporal and spatial variations in these sediments in order to assess their sources.

Four gravity cores were taken in 2011 (Fig. 6.21). Core 5 was the longest (Fig. 6.22) and acts as a key core. Dried core sections (2 cm intervals) were subjected to a suite of analyses:

- pXRF analysis (Niton XL3t 900);
- mineral magnetic characterization (Bartington Instruments Ltd. MS2 and C sensor, Molspin Ltd. Pulse Magnetiser and Minispin Fluxgate Magnetometer);
- total organic matter (550 °C) and carbonate content (850 °C) determination via LOI particle size measurement (Horiba Partica LA-950V2).

Core 5 provides an overview of the properties of the recent sediments (Fig. 6.23). A major change at *c.*40 cm (e.g., LOI, particle size, magnetic properties and geochemistry) indicates an interface between a possible flood deposit and the

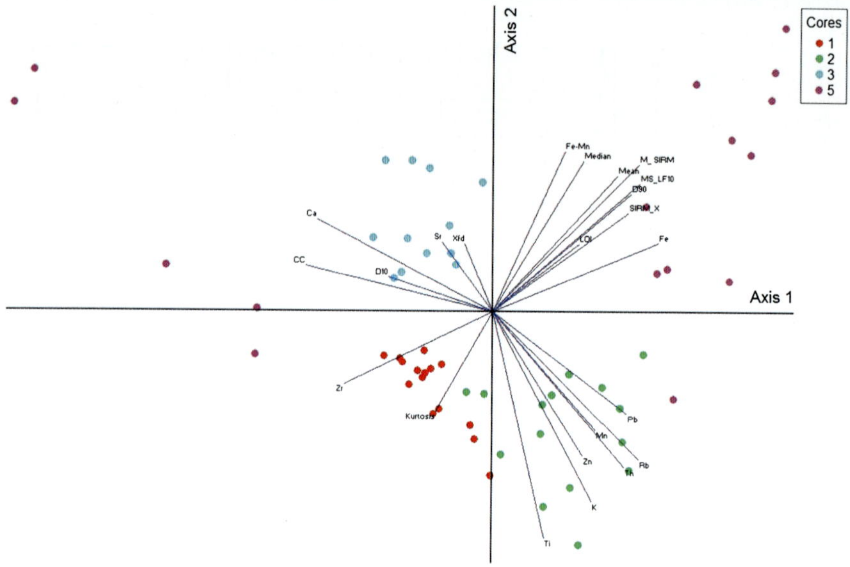

Fig. 6.24 PCA biplot of components 1 and 2. Samples from the base of core 5 cluster towards the top left and are clearly differentiated from the rest of this core (with a contrasting geochemistry (e.g., Ca vs. Fe)), physical properties (e.g., carbonate content vs. particle size characteristics) and magnetic concentrations. The remaining sediment samples are clearly grouped and core 3 is also discriminated (particularly by its Xfd.)

overlying lacustrine sediments. Subsequently, the relative importance of surface sources of sediment (shown by *X*fd) fluctuates and there is a marked change in the particle size distribution at *c.*17 cm, which may be approximately dated to the mid 1980s by the Cs-137 levels (Hutchinson et al. 2012). The Fe/Mn ratio also highlights the marked change in the sediment's properties mid core. It suggests that the site has recorded a number of significant sedimentological changes in addition to the post 1989 (logging) impacts previously reported (e.g., Begy et al. 2009).

PCA has been used to interrogate the entire data set and highlights spatial variations in the sediments; cores along the main limb of the lake share common characteristics, whereas core 3 (Suhard brook) demonstrates contrasting properties (particularly LOI). Furthermore, cores 1, 2 and 5 do not become finer from the head to the main body of the lake. Both features may reflect human intervention in the catchment and the lake i.e., forestry work close to the lake on the Suhard brook and the construction of sediment lodges from the 1960s on the Sheep and Red brooks (Fig. 6.24) (Hutchinson et al. 2012).

Preliminary data suggest that the recent sediments of the Red Lake can be readily characterized highlighting different sediment zones and thereby the changing relative importance of sediment sources both temporally and spatially (Hutchinson et al. 2012). This information can contribute to environmental

Fig. 6.25 Two views of the Red Lake in the 1930s and mid 2000s. Note the nature and extent of the forest cover

management decision making for the site. Furthermore, the palaeolimnological record of the lake offers an important perspective for the assessment of the impacts of forest management that is a regionally significant issue (Fig. 6.25).

References

Begy R, Cosma C, Timar A (2009) Recent changes in Red Lake (Romania) sedimentation rate determined from depth profile of ^{210}Pb and ^{137}Cs radioisotopes. J Environ Radioact 100:644–648

Bojoi I (1968a) Observaţii asupra reliefului carstic din masivul Hăghimaş. In: Proceedings of the station for the biologic, geologic and geographic research Stejaru, vol 1

Bojoi I (1968b) Contributii la sedimentologia Lacului Rosu. In: Proceedings of the station for the biologic, geologic and geographic research Stejaru (in Romanian) vol 1. pp 87–105

Clague J, Luckman BH, Van Dorp RD, Gilbert R, Froese D, Jensen BJL, Reyes AV (2006) Rapid changes in the level of Klaune lake in Yukon territory over the last millennium. Quat Res, Elsevier 66:342–355

Forrest AL, Laval BE, Pieters R, Lim DSS (2008) Convectively driven transport in temperate lakes. Limnol Oceanogr 53(5):2321–2332

Hutchinson SM, Armitage R, Frantiuc A, Haliuc A, Mîndrescu M, Romanescu G (2012) The recent palaeolimnology of Red lake (Romania); assessing sediment sources for environmental management. Symposium: water resources and wetlands, 14–16 Sept 2012, Tulcea, Romania (Poster)

Miller MP, Mcknight DM, Chapra SC, Williams MW (2009) A model of degradation and production of three pools of dissolved organic matter in an alpine lake. Limnol Oceanogr 54(6):2213–2227

Mulholland PJ, Kuenzler EJ (1979) Organic carbon export from upland and forested wetland watersheds. Limnol Oceanogr 24:960–966

Nechita N (2003) Flora si vegetatia cormofitelor din masivul Hasmas, Cheile Bicazului si Lacului Rosu (The flora and vegetation of the cormophytes from Hasmas Massif, Bicaz Gorges, and the Red Lake). Bibliotheca Historiae Naturalis II, Natural Sciences Museum, Piatra Neamt, pp 1–383

Pandi G (2004) A Gyilkos-Tó. Hidrogeográfiai tanulmány. Casa Cartii de Stiinta Publishing House, Cluj, pp 1–231

Pişotă I, Năstase A (1957) Lacul Rosu, nod de confluenta a trei bazine hidrografice. Probleme de geografie (in Romanian) 4:181–205

Romanescu G, Cojocaru I (2010) Hydrogeological considerations on the western sector of the Danube Delta—a case study for the Caraorman and Saraturile fluvial-marine levees (with similarities for the Letea levee). Environ Eng Manage J 9(6):795–806

Strahler AN (1952) Hypsometric (area-altitude) analysis of erosional topography. Bull Geol Soc Am 63:1117–1142

Chapter 7
Climatic Setting

Abstract The area of the Red Lake and the Hăghimaş Mountains is included in a mountain thermal climate, with great diurnal and seasonal variations. There are no meteorological and hydrologic posts in the area of the Red Lake hydrographic basin and data were taken from expeditionary measurements made by Ghenciu (Yearbook Museum Nat Sci Piatra Neamţ, Geol-Geogr, 3:313–319, 1976) in 1968–1970 or they came from interpolation (Apăvăloaie 1971, 1980; Mihăilescu in Observaţii microclimatice expediţionare în uvala Terkö (Piatra Crăpată), Pângăraţi, p 7, 1980 and Apăvăloaie 1980). The specific winds are mountain-valley breezes: from the peak towards the valley in the evenings and the other way around during the day. The big difference in the relative altitude makes the breezes frequent and relatively strong. In the evenings, during the mountain-valley breezes, a strong smell of resin—from the high-altitude pines—fills the air. The relatively cold climate is determined by the appearance of thermal inversions. They are frequent in the winter and they make the ice bridge that occurs on the lake, very long-lasting. Though the Red Lake is situated in a mountain area, there have been examples of temperatures above freezing in winter (4.3 °C in February 1968, 1.1 °C in December 1970, or 0.4 °C in December 1969). For a mountain area, with frequent thermal changes, especially in the winter, the values are not surprising. The climate of the Red Lake area is typical for the surroundings and it is a representative example of the mountainous climate of the Oriental Carpathian mountain range.

Because there are no meteorological and hydrologic posts in the area of the Red Lake hydrographic basin, the data regarding the climatic phenomena have only a general character. Some of them were taken from the expeditionary measurements made by Ghenciu (1976), in 1968–1970, or they came from interpolation (Apăvăloaie 1971, 1980; Mihăilescu and Apăvăloaie 1980). The area of the Red Lake and the Hăghimaş Mountains is included in the mountain thermal climate, with great diurnal and seasonal variations.

The atmospheric pressure means and the concentration of oxygen in the atmosphere are relatively low (about 18 % than that of the surrounding area).

G. Romanescu et al., *Limnology of the Red Lake, Romania*,
DOI: 10.1007/978-94-007-6757-7_7,

Though the breathing rhythms decreases, its deepness makes us inhale a larger amount of air. This way, the blood cells are better oxygenated and cardiac activity is optimal. The large amount of oxygen makes the burning within tissues and the nutritive exchanges complete. As a result, blood pressure decreases, balancing the nervous system and reducing stress. From this perspective, the area of the Red Lake has been exploited for health-related and climatic tourism, making it a high-potential resort.

Because of the depression-like aspect, the winds are mild; most of the times, they change in direction and intensity. The specific winds are mountain-valley breezes: from the peak towards the valley in the evenings and the other way around during the day. The big difference in the relative altitude makes the breezes frequent and relatively strong. In the evenings, during the mountain-valley breezes, a strong smell of resin—from the high-altitude pines—fills the air.

A foehn can be perceived on the slopes covered from the air masses coming from the west. This is why there is more precipitation in front of the western winds, climbing the slopes and less precipitation on the eastern slopes, where they go downwards. The mean annual thermal amplitude varies from 22 to 24 °C. The minimum average temperature oscillates between −7 and −5 °C in January, while the maximum average temperature reaches 17–18 °C in July. The annual thermal mean varies between 5 and 7 °C, depending on the altitude. There is a long period with low temperatures, the reason for which the solid precipitations fall on a long-term basis; the mean is over 120 days per year.

The month with the lowest temperature is February, with −7 °C and the highest values are in August −15 °C (Table 7.1). The average monthly temperature with the lowest value was −9.4 °C in January 1969 and the highest was 14.6 °C, in June 1968. There are significant temperature differences between the winter of 1969, more frosty and that of 1970, less chilly (Table 7.2).

Though the Red Lake is situated in a mountain area, there have also been temperatures above freezing in winter (4.3 °C in February 1968, 1.1 °C in December 1970, or 0.4 °C in December 1969). The very high temperature value of

Table 7.1 The average monthly temperature within the area of the Red Lake hydrographic basin, for 1950–1975, based on interpolations

Month	I	II	III	IV	V	VI	VII	VIII	IX	X	XI	XII
°C	−5	−7	−1	3	8	12	14	15	11	6	1	−3

Table 7.2 The average monthly temperature within the area of the Red Lake basin in 1968–1970 (°C) (Ghenciu 1976)

Year	Month											
	I	II	III	IV	V	VI	VII	VIII	IX	X	XI	XII
1968	–	–	–	–	12.1	14.6	14.3	13.4	10.8	5.2	2.4	−5.6
1969	−9.4	−5.4	−4.8	2.1	11.6	8.6	13.1	13.4	9.5	4.2	3.4	−4.5
1970	−4.3	−4.3	−1.5	5.0	8.3	13.1	15.1	13.7	9.6	3.9	2.2	−2.6

Table 7.3 The average maximum daily temperature within the area of the Red Lake basin in 1968–1970 (Ghenciu 1976)

Year	Month											
	I	II	III	IV	V	VI	VII	VIII	IX	X	XI	XII
1968	−4.1	4.3	4.4	17.2	16.0	20.2	19.2	18.8	13.3	12.6	7.2	−1.0
1969	−2.4	−1.3	0.8	7.6	18.2	17.0	17.6	22.8	18.1	13.7	10.2	0.4
1970	0.9	0.7	3.3	9.8	13.4	18.1	21.8	19.9	15.7	9.6	5.9	1.1
Interpolation	−3.0	−3.0	3.0	10.0	14.0	18.0	18.0	17.0	14.0	9.2	7.2	−2.0

the month of April 1968 −17.2 °C should be underlined. In the year 1970, there were only positive values (Table 7.3). The average minimum daily air temperatures vary between −15.2 °C in January 1969 and 9.8 °C in July 1970. For a mountain area, with frequent thermal changes, especially in the winter, the values are not surprising (Table 7.4).

The absolute maximum temperatures may be positive even during the winter. There are also absolute minimum temperatures, below freezing, in the transition months, meaning during spring and autumn. There may also be 1 °C in the summer (July 1968) and 28 °C in the spring (April 1968). The mean monthly thermal amplitude was 32.9 °C in 1968, 38 °C in 1969 and 22.7 °C in 1970. The absolute thermal amplitude was 58.5 °C in 1968, 54.6 °C in 1969 and 49.5 °C in 1970 (Table 7.5).

The sums of the average daily temperatures underline the repartition method on months and seasons. The energy increases gradually, since winter until spring, with the maximum value in the summer and then decreasing again during the cold season (Table 7.6). As regards the sum of the average daily temperatures, there are big differences between August and September–October. In the month of June there was a low value −258.1 °C in the year 1969 (Table 7.7).

Overall, the winters are cold, because of the continental air masses coming from the east. Often, there are temperatures of −25 and −30 °C. Due to altitudinal layering, summers are rather cool, with a mean of 18 °C in July. The first frost usually appears at the beginning of October and the last one in the second 10-day period of May.

The relatively cold climate is also determined by the appearance of thermal inversions. They are frequent in the winter and they make the ice bridge very long-

Table 7.4 The average minimum daily temperature within the area of the Red Lake basin in 1968–1970 (Ghenciu 1976)

Year	Month											
	I	II	III	IV	V	VI	VII	VIII	IX	X	XI	XII
1968	−12.1	−9.0	−9.5	−0.6	5.1	6.2	8.4	6.9	7.7	0.4	−1.5	−11.0
1969	−15.2	−10.4	−9.7	−2.6	4.2	6.6	8.2	7.9	3.3	−1.1	−1.8	−7.2
1970	−9.0	−9.2	−6.2	−2.0	4.0	6.6	9.8	7.8	4.3	−1.3	−1.3	−6.1
Interpolation	−11.0	−9.0	−7.0	−10.0	3.0	7.0	7.0	7.0	5.0	−3.0	−3.0	−7.0

Table 7.5 The absolute maximum and minimum temperatures within the area of the Red Lake basin between 1968 and 1970 (Ghenciu 1976)

Month	Absolute maximum temperature			Absolute minimum temperature		
	1968	1969	1970	1968	1969	1970
I	4.5	3.0	11.5	−29.0	−25.0	−22.0
II	13.0	8.0	9.0	−17.0	−22.0	−21.0
III	19.4	11.0	12.0	−24.3	−25.0	−17.0
IV	28.0	18.0	20.0	−7.0	−10.0	−9.0
V	26.0	26.5	23.0	0	−3.0	−1.0
VI	25.5	23.0	23.0	1.0	1.5	4.0
VII	29.5	23.5	27.5	4.0	3.0	5.0
VIII	24.0	29.5	25.0	2.0	3.5	3.0
IX	26.0	26.0	26.5	−1.0	−4.5	−2.0
X	22.5	21.5	19.5	−8.5	−4.5	−9.0
XI	19.0	20.0	15.0	−8.5	−13.0	−7.0
XII	7.5	10.0	6.0	−20.0	−13.5	−12.0

Table 7.6 Sum of the average daily temperatures on 10-day periods in the Red Lake area between 1968 and 1970 (Ghenciu, 1976)

Month	10-day period								
	1968			1969			1970		
	I	II	III	I	II	III	I	II	III
I	–	–	–	−122.8	−73.8	−81.3	−40.5	−13.4	−76.9
II	–	–	–	−64.8	−54.4	−45.6	−19.3	−65.6	−36.9
III	–	–	–	−72.1	−32.2	−48.7	−29.0	−32.0	12.6
IV	–	–	–	1.1	19.6	54.6	26.4	30.6	99.7
V	–	–	–	103.6	139.5	116.3	70.3	88.0	100.0
VI	138.5	149.6	150.7	94.6	137.2	126.3	100.0	138.0	147.0
VII	164.5	154.4	126.8	137.7	120.4	135.2	158.0	146.0	168.0
VIII	132.9	136.4	145.3	125.0	148.0	144.0	163.0	133.0	129.0
IX	125.0	122.0	76.8	108.2	96.0	78.9	108.0	146.0	36.0
X	53.2	73.0	37.7	52.1	36.3	41.6	63.0	26.0	34.0
XI	60.3	13.7	−1.1	25.9	53.2	25.7	40.0	24.3	4.1
XII	−57.0	−60.7	−6.8	−12.6	−49.4	−78.0	−15.8	−47.5	−20.0

lasting (Fig. 7.1). The sun is up longer in August and September and shorter in November and December. The annual mean sums up to 1,800 h. The high values of UV radiations facilitate the heliotherapy and aerotherapy sessions within the depression area of the lake.

The rainfalls depend on air masses and altitudinal layering. The western slopes are rainier than the eastern ones. The annual mean of rainfalls is over 600 mm. The rainy period starts in the second part of the month of May and it ends at the beginning of July. The lowest amount of rainfalls is registered in the months of August and September or in February and March.

Table 7.7 Sum of the average daily temperatures within the area of the Red Lake in 1968–1970 (Ghenciu, 1976)

Month	Year		
	1968	1969	1970
I	–	–278	–130.8
II	–	–168.8	–121.8
III	–	–153.0	48.4
IV	–	75.3	156.7
V	–	359.6	258.3
VI	438.8	258.1	395.0
VII	445.5	405.8	472.0
VIII	414.6	417.0	425.0
IX	323.8	286.1	290.0
X	172.9	130.0	123.0
XI	72.0	104.8	63.4
XII	–174.5	–138.9	–83.3

Fig. 7.1 Ice bridge in the winter of 2010

During the pluviometric maximum period, the rains are heavy and sometimes long lasting. Solid precipitations represent 20–23 % of the total. The snow cover in the depression areas lasts over three months (Apăvăloaie 1971, 1980). Towards the end of winter, the snow layer may reach up to 1–1.20 m on the western slope and to 0.70–0.90 m on the eastern one. The air's relative humidity is 65–75 % in the summer and 80–85 % in the winter. From the bioclimatic perspective, these values are normal.

The bioclimatic value is amplified by the high concentration of negative oxygen ions, due to the intense evaporation on the surface of the lake, the fragmentation of water drops formed after the creation of small waterfalls, UV radiation and terpenic (resin) aerosols from the pine forests. The negative oxygen ions are also known as "air vitamins" as they increase man's vitality force, resistance to disease, longevity, etc.

The vegetations and UV radiations make the level of ozone concentration in the air high. This way, the air is purified and the blood circulation is stimulated, blood pressure is reduced, inflammatory processes are diminished and appetite is stimulated. The iodine concentration in the atmosphere is higher than in the surrounding regions and it is conjugated with the positive effects of the other elements. The potholes on the Valea Seacă, with temporary waters, only during rainfalls, demonstrate the existence of a climate with high values of rainfalls in the older times of the Red Lake basin.

The "L" shape of the Red Lake basin makes the two arteries continue into the valleys of the Oaia (Oii) and Suhard brooks. From this point of view, three topoclimatic areas are delimited: the lower stream of the Oaia (Oii) brook, the lower stream of the Suhard brook and the surface of the Red Lake.

The topoclimates present the following characteristics:

- the summer temperatures are higher on the Suhard stream;
- the winter and spring temperatures are lower on the Suhard stream;
- the ice bridge is longer lasting on the Suhard stream, which is also in the shade for a longer period, due to its east-westward orientation (in the spring and autumn);
- the wind is stronger and more frequent on the Oaia (Oii) stream;
- the atmospheric calmness is specific to the Suhard stream;
- the sun shines for a longer period on the Oaia (Oii) stream, due to its south-northward orientation;
- the sun shines for a longer period, during the summer, on the Suhard stream;
- the sun shines for a longer period, during the winter, on the Oaia (Oii) stream;
- on the surface, mostly at the confluence of the two arteries, the conditions are intermediary.

References

Apăvăloaie M (1971) Contribuţii la cunoaşterea regimului temperaturii aerului din zona Lacului Roşu-Cheile Bicazului. Lucrările Staţiunii de Cercetări Stejaru, pp 227–237

Apăvăloaie M (1980) Contribuţii la cunoaşterea distribuţiei precipitaţiilor atmosferice din bazinul râului Bistricioara. Lucrările Staţiunii de Cercetare Stejaru, Geologie-Geografie 7:88–94

Ghenciu IV (1976) Temperatura aerului şi regimul ei din zona Lacului Roşu. Yearbook of the Museum for Natural Sciences Piatra Neamţ, Geology-Geography 3:313–319

Mihăilescu IF, Apăvăloaie M (1980) Observaţii microclimatice expediţionare în uvala Terkö (Piatra Crăpată). In: Proceedings of the Stejaru Station, Geology-Geography, Pângăraţi, 7

Chapter 8
The Seasonal Variation of Temperature, pH and Dissolved Oxygen Concentration

Abstract In order to have a complete picture of the limno-ecological conditions, 17 measuring and sampling points were chosen. The points chosen covered the whole of the bathymetric and morphological spectrum of the lake. The measurements were carried out on the surface and from meter to meter as far as the bottom and they were repeated in every season; spring, summer, autumn and winter. The pH of the water is found to be slightly alkaline throughout the year with a range of 7.5–8.0 (Fig. 8.5). Very rarely, in certain conditions, the pH value falls within the circum-neutral category. The existence of dissolved oxygen is essential for the water table. Even given this, the Red Lake is not the object of extraordinary biological diversity, which follows from the fact that its existence is relatively recent. All these parameters are important because, at their junction, they define the vegetal and animal life conditions that the Red Lake has to offer. The geographic location of the lakes in Romania, in distinct landform units (mountains, hills, and plains, littoral) creates different environments for the manifestation of trophicity. According to the value of the nutrients, the greatest parts of the lakes (wetlands) are situated in the mesotrophic, meso-eutrophic, eutrophic, eutro-hypertrophic and hypertrophic categories.

In order to have a complete picture of the limno-ecological conditions, 17 measuring and sampling points were chosen. The points chosen covered the whole of the bathymetric and morphological spectrum of the lake. The Hach Lange Multiparameter was used for a rapid analysis of the physical and geographic factors. The measurements were carried out on the surface and from meter to meter as far as the bottom. All the measurements were repeated in every season; spring, summer, autumn and winter. This way, sufficient data were obtained for the correct analysis of the variation in living conditions of the water table. At the same time, the sampling points were chosen especially in order to register the variations for water from streams.

The positioning of the sampling points was carried out with the help of GPS. Measurements at the same point could not always be carried out because of the existence of aquatic currents on the surface but more especially of those at the

G. Romanescu et al., *Limnology of the Red Lake, Romania*,
DOI: 10.1007/978-94-007-6757-7_8,

bottom. For this reason, seasonal measurements did not always attain the same bathymetric values.

The sampling took place in the daytime between 10 AM and 4 PM. The level variations in the lake maintain their maximum values of 60–70 cm, with low values in the winter and high values in the spring, after snow melting (Ghenciu 1972). Water transparency, measured with a Secchi disc, varies between 0.20 and 3.5 m. The mean is around 2–2.3 m. The minimum transparency is registered during heavy rains and in the spring, when snow melts. The maximum transparency is registered in the winter and at the end of autumn, when a part of the plankton disappears.

During high waters, the water becomes yellowish and in the rest of the year, it is greenish. The red color appears only at sunrise, when the deposits full of iron oxides are reflected in the lake waters (Fig. 8.1). The thermal values of the pH and the dissolved oxygen were registered for the entire vertical section of the lake basin: from the surface to a maximum depth of 10.5 m.

The most reduced depths are to be found at the two extremities, where the Suhard and Oii streams empty, where formations of the fan-delta type form. Maximum depths are to be found at the river mouth, where the present Bicaz River is formed. In reality, the sector with the greatest depths corresponds to the confluence between the two brooks, a favorable location for the formation of one eddy current that agitates the bottom of the lake basin. Because of their origins, the lateral sections of the "bowl" present depths that are still high from the bank, especially on the rocky, steep, western slopes.

On the surface, the maximum temperature is 18 °C and continues until autumn. In the summer, a maximum temperature of 17.8 °C has been registered. The minimum temperature in wintertime favors the formation of ice bridges that can

Fig. 8.1 The abrupt slope with red limestones reflecting in the lake waters

reach dimensions of 20–70 cm. In the winter of 2009, the maximum dimension was 24 cm. In spring, a minimum temperature of 4 °C was registered. The ice bridge may last 120–130 days (Figs. 8.2, 8.3).

At a depth of 10 m the maximum temperature is 6 °C, with a minimum of 4 °C. The temperatures are maintained between 4 and 6 °C. From a thermal point of view, the Red Lake behaves like a dimictic water body, with direct thermal stratification in the summer and the opposite in winter. On the surface, the pH values rise to a maximum of 8.5 in springtime. The minimum falls to 6.5 in wintertime. The values are usually maintained at between 7 and 8.2 (Fig. 8.4).

At a depth of 10 m a maximum of 7.6 is maintained in the autumn and a minimum of 7.2 in the springtime. The values are maintained at 7.2–7.8. The quantity of dissolved oxygen from the waters of the lake depends on temperature, circulation, water quality, etc. (Fig. 8.5). The decomposition of organic remains within the lacustrine basin has a great influence upon the quantity of the dissolved oxygen (Sobek et al. 2009).

On the surface, a maximum of 10.6 mg/L has been registered in wintertime. The minimum produced in summer is 7 mg/L. The values for spring and autumn are intermediate, between 7 and 10 mg/L. Lack of dissolved oxygen can be perceived beginning at a depth of 8 m. At the bottom of the lake, dissolved oxygen is lacking.

The fact that the Red Lake is located inside a mountainous depression of small dimensions, almost completely closed, favors the frequent appearance of thermal opposites. This is the reason why the values of water temperatures do not connect correctly with the air temperatures at the meteorological stations nearby. Unfortunately, there is no meteorological station in the mountainous depression to make the correlations between the mean water value and water temperature values.

Fig. 8.2 Measurement of the ice bridge in the winter of the year 2010

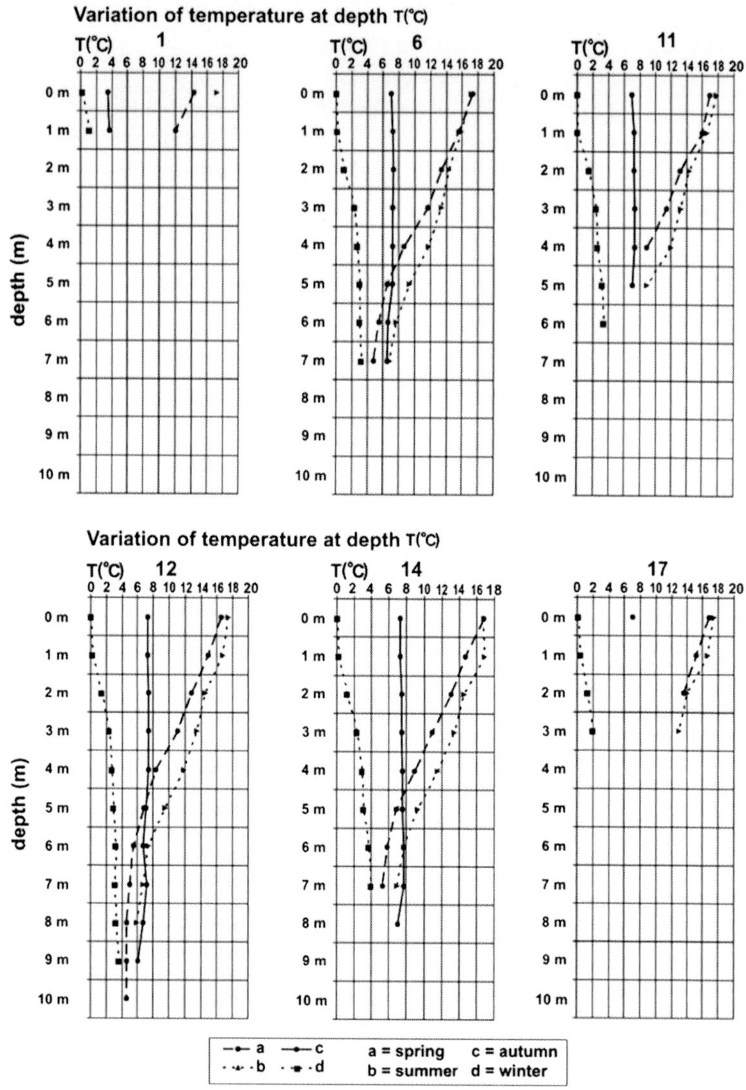

Fig. 8.3 Seasonal distribution (*vertical*) of thermal values in the Red Lake (sample characteristics)

On the surface, the maximum temperatures are achieved in the spring and summer. These are specific to the central zones, where there are no lateral water tributaries originating from the brook (Fig. 8.6a, b). Temperatures in summertime and spring follow this trend and are close. The maximum temperature of the water is a few degrees (2–3 °C) lower than that in the air.

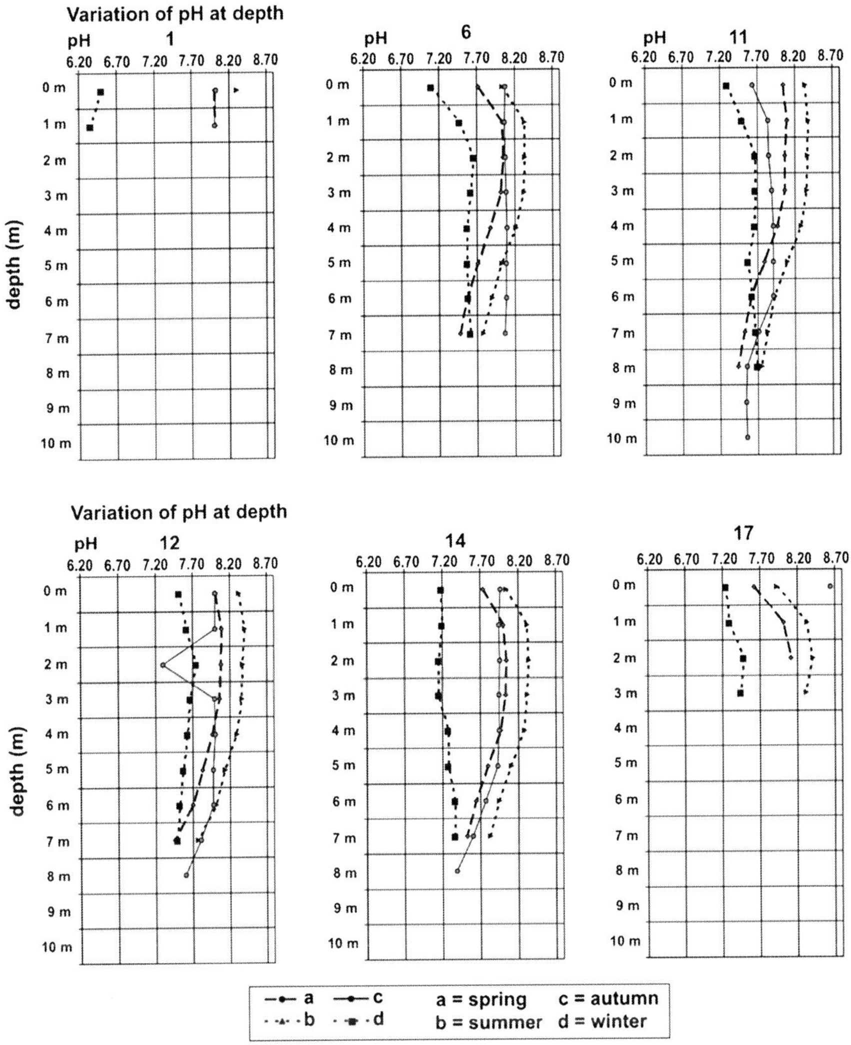

Fig. 8.4 Seasonal distribution (*vertical*) of the pH in the waters of the Red Lake (sample characteristics)

In the autumn, temperatures fluctuate between 6 and 8 °C (Fig. 8.6c). In wintertime, bridges of ice are formed (Fig. 8.6d). Bridges of ice can reach dimensions of up to 70 cm. The dimensions of the ice bridges are usually uniform on the whole surface of the lake. This fact is favored by the existence of fossilized tree trunks spread across almost the entire surface of the lake.

In the summer, the temperature falls relatively easily to a depth of 3.5 m. The thermal homogenization of the water determines the existence of some relatively

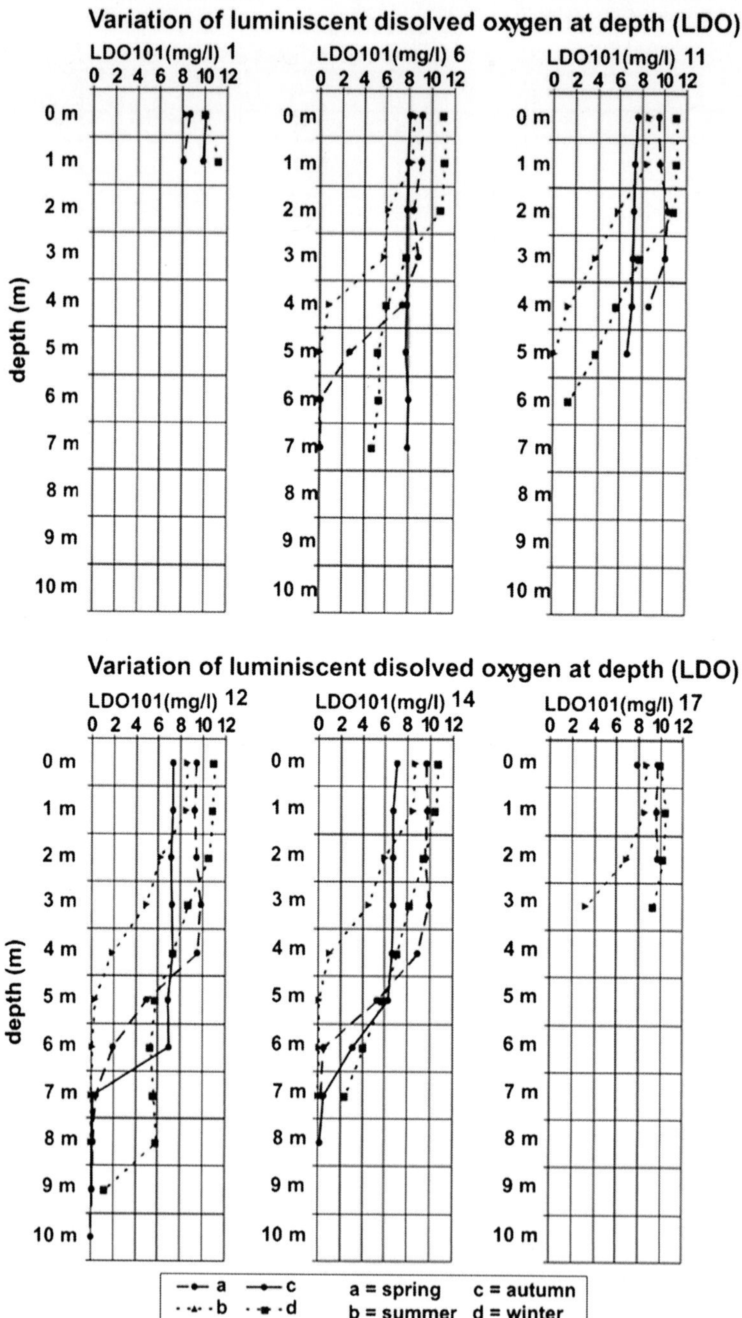

Fig. 8.5 Seasonal distribution (*vertical*) of dissolved oxygen in the Red Lake (sample characteristics)

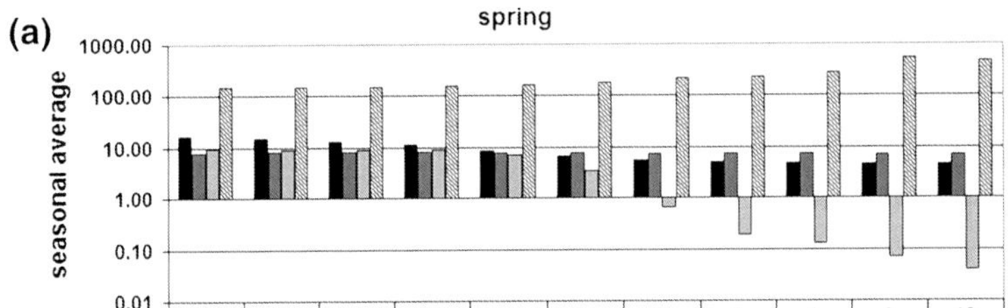

	0	1	2	3	4	5	6	7	8	9	10
■ T (°C)	16.78	15.04	13.35	11.33	8.67	6.82	5.68	5.04	4.90	4.60	4.60
■ pH	7.99	8.12	8.11	8.06	7.93	7.76	7.62	7.50	7.52	7.26	7.26
□ LDO (mg/L)	9.27	9.21	9.00	8.84	7.14	3.53	0.67	0.19	0.13	0.07	0.04
▧ CDC401(mg/L)	150.43	151.15	152.28	155.42	160.97	182.74	214.27	228.84	281.03	528.00	476.00

depth (meters)

Fig. 8.6 **a** Vertical distribution of temperatures, pH and dissolved oxygen in springtime, **b** Vertical distribution of temperatures, pH and dissolved oxygen in summertime, **c** Vertical distribution of temperature, pH and dissolved oxygen in autumntime, **d** Vertical distribution of temperatures, pH and dissolved oxygen in wintertime

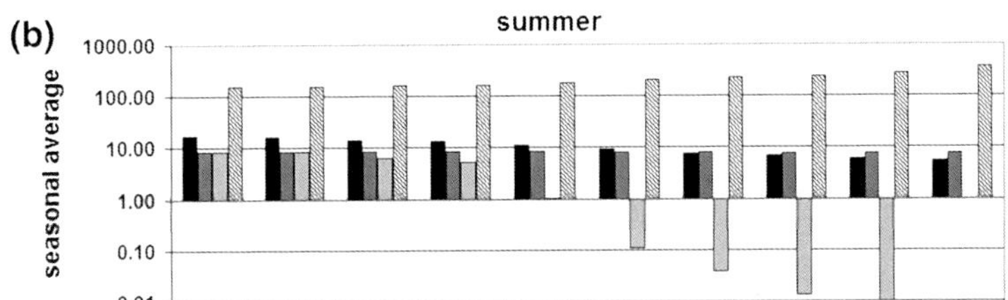

	0	1	2	3	4	5	6	7	8	9
■ T (°C)	17.13	16.16	14.37	13.35	11.45	9.15	7.71	6.81	6.04	5.53
■ pH	8.27	8.42	8.42	8.36	8.25	8.08	7.97	7.84	7.77	7.81
□ LDO (mg/L)	8.49	8.28	6.33	5.25	1.04	0.11	0.04	0.01	0.01	
▧ CDC401 (mg/L)	156.32	155.30	161.53	162.10	176.35	201.91	222.76	239.80	271.68	362.82

depth (meters)

Fig. 8.6 (continued)

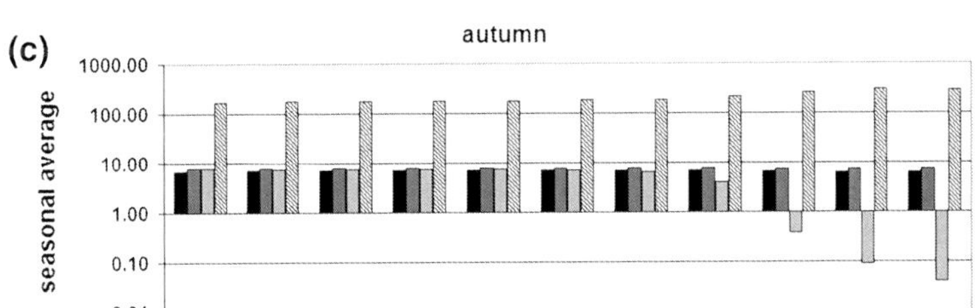

	0	1	2	3	4	5	6	7	8	9	10
■ T (°C)	6.81	7.01	7.26	7.29	7.25	7.30	7.01	7.02	6.97	6.35	6.40
■ pH	8.04	8.02	7.98	8.02	7.99	8.05	8.01	7.89	7.62	7.57	7.55
□ LDO (mg/L)	7.83	7.60	7.48	7.42	7.34	7.08	6.35	3.97	0.38	0.09	0.04
▧ CDC401 (mg/L)	171.72	181.31	181.94	181.85	181.67	182.63	188.81	210.98	266.72	298.83	290.00

depth (meters)

Fig. 8.6 (continued)

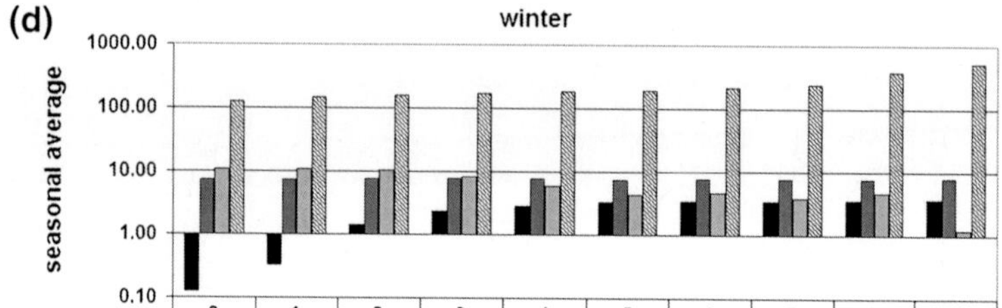

depth (meters)	0	1	2	3	4	5	6	7	8	9
■ T (°C)	0.13	0.33	1.44	2.36	2.83	3.25	3.36	3.45	3.57	3.72
■ pH	7.33	7.45	7.62	7.57	7.57	7.41	7.61	7.60	7.57	7.83
□ LDO (mg/L)	10.98	10.97	10.20	8.17	5.90	4.35	4.69	3.82	4.70	1.23
▨ CDC401 (mg/L)	124.44	147.66	157.28	173.34	181.72	192.80	214.60	245.24	382.09	530.53

depth (meters)

Fig. 8.6 (continued)

uniform temperatures. At this depth, it reaches 13–14 °C. Below these depths, the temperature drops rapidly and reaches minimum values of 6 °C at the greatest depths. In the spring, this tendency is respected. The only difference lies in the fact that the temperatures are lower by approximately 1–2 °C than those in summer.

An extremely interesting fact features in the vertical homogenization of temperatures in autumn. These temperatures maintain values of 6–7 °C at the surface. At depths of more than 10 m they reach a maximum of 8 °C. Homogenization can be the result of the existence of the greater flow rate of brooks during autumn, the result of some local winds that come down from the slopes and the result of the disappearance of vegetation, which makes the mass of water have almost perfect homogeneity.

In winter, heat homogeneity is lower and it shows on a scale of 4 °C (between 0 and 4 °C). Up to a depth of 1 m the temperatures remain at 0 °C as a result of the influence represented by the ice bridge. At the depth of 3.5 m temperatures rise relatively quickly to 2–2.5 °C. Between 3.5 m depth and the bottom of the lake (10.5 m) temperatures rise to 4 °C.

Thus, the phenomenon of thermal inversion appears to be more obvious in winter and less intense in autumn. In summer and spring, the normal phenomenon of lower values varies by depth. At the surface, the pH value varied between 6.5 and 8.5. In winter and springtime, the values are almost identical, around 8. The autumn values also remain at around 8. Nonetheless, in wintertime the values are much lower (6.4–7.5). A slight change on the vertical of about 5–6 m can be noticed in summer, autumn and springtime, from which the pH goes down relatively brusquely from 8 to 7.5. The greatest depth of the pH maintains the value of 7.5.

The only difference seen, compared to other temperatures, is at wintertime. The pH values in wintertime are maintained at around 7. If, for the other three seasons, the pH decreases slightly with depth, in the winter it grows, reaching 7.5. The only value of 8 is reached at sampling point 12, where there are great depths and the two currents come together and determine the existence of the Suhard and Oii streams.

The pH of the water is found to be slightly alkaline throughout the year with a range of 7.5–8.0 (Fig. 8.7). Very rarely, in certain conditions, the pH value falls within the circum-neutral category. The existence of dissolved oxygen is essential for the water table. Even given this, the Red Lake is not the object of extraordinary biological diversity, which follows from the fact that its existence is relatively recent.

The greatest values of dissolved oxygen are registered in the winter, when water, at its coldest, produces the most important distortions of the thresholds. In the summer, the values are reduced, although they are sufficient for the maintenance of optimum ecologic conditions. They are values specific to mountainous waters, especially rivers. The high value of dissolved oxygen is due to the streams feeding the water body. This gives rise to currents that clear the surface water.

Between summer and winter, there is an important difference of values once depth increases. In the summer, at a depth of 4 m, the values fall beneath 4 %. In the winter, these values fall at a depth of 7–8 m. For certain intermediary seasons (spring and autumn), the lowest values begin at a depth of 6 m.

In the sectors with the greatest depths and where the meeting of two currents produce two streams (Suhard and Oaia) the most important distortion of the threshold limits occur. In this case, the thresholds are greatly expanded to between 4 and 7 m. It should be noted that in this sector the waters are being continuously agitated (with the exception of the appearance of ice bridges) by the oars of tourist boats that provide boat rides in this area.

Throughout the whole year, a good correlation with the three values can be observed. They drop from the surface to depths in three seasons: spring, summer and autumn (Fig. 8.8). These values are unbalanced in wintertime, especially by the presence of ice bridges that form every year and can remain for between 20–80 days per year. The extremely long existence of the ice bridges is due to the

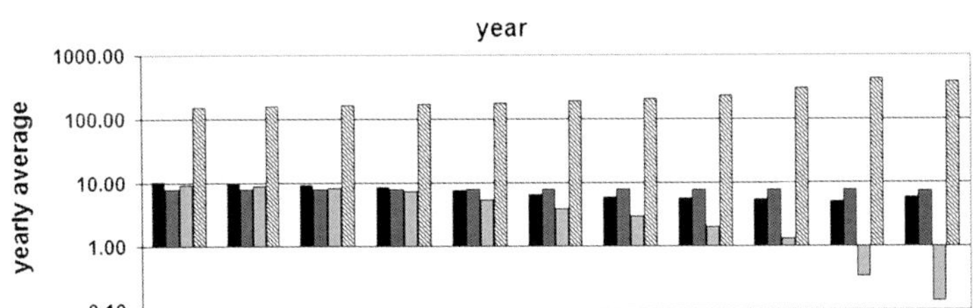

	0	1	2	3	4	5	6	7	8	9	10
■ T (°C)	10.21	9.64	9.10	8.58	7.55	6.63	5.94	5.58	5.37	5.05	5.78
■ pH	7.91	8.00	8.03	8.00	7.94	7.82	7.80	7.71	7.62	7.62	7.41
□ LDO (mg/L)	9.14	9.01	8.25	7.42	5.36	3.77	2.94	2.00	1.30	0.35	0.14
▨ CDC401 (mg/L)	150.73	158.85	163.26	168.18	175.18	190.02	210.11	231.21	300.38	430.04	383.00

depth (meters)

Fig. 8.7 Vertical distribution of annual temperature average, pH and dissolved oxygen in the Red Lake

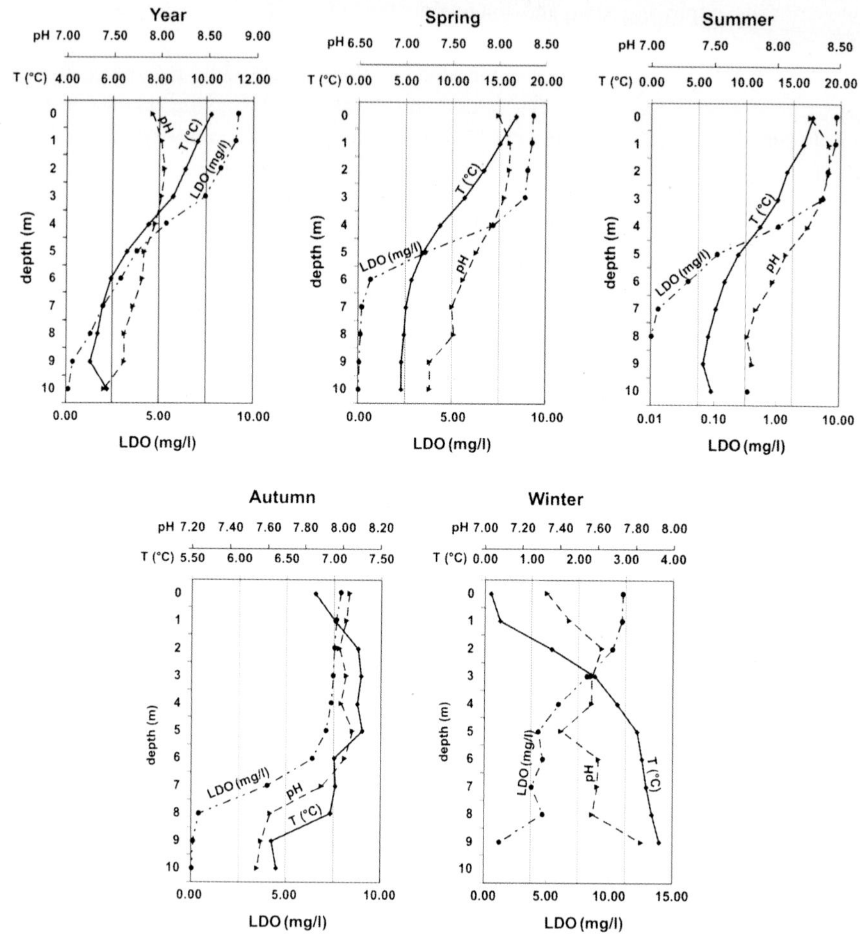

Fig. 8.8 Annual variation and by season of temperatures, pH and dissolved oxygen in the waters of the Red Lake

formation of inverse thermal conditions in winter that favor the installation of a table of cold air at the bottom of the depression.

Ice bridges are relatively uniform, with a size proportional to the entire surface of the lake and with values that can vary between 15 and 70 cm, the most frequent being those limited to 20 cm. The size depends on the thermal values of the air and their duration. They are larger in years in which thermal inversions take place, which lead the minimum temperatures to −37 °C.

Water quality depends on all the physical, chemical and bacteriologic characteristics that influence the environment. The quality of the lake waters is determined by the presence of all mineral and organic substances, dissolved gases, particles in suspension and living organisms. The impurities are dispersed in water as suspensions, colloids and solutions.

Water conductivity, besides the total dissolved salts (TDS), constitutes the most frequently used indicator to assess the mineralization degree of the waters. If there is a high concentration of salts, water conductivity will also be bigger. Only pure water does not conduct electricity. The measurements regarding water conductivity were taken in the same points and at the same time as those for temperature, pH and dissolved oxygen.

The Red Lake waters have normal conductivity values, as well as normal TDS values: 150–300 mg/L (Figs. 8.9, 8.10, 8.11, and 8.12). There were maximum values of 490–530 mg/L in the sampling points 7 and 12 and minimum values of 120–130 mg/L in the points 11, 14 and 16.

The total dissolved salts have a lower value in the epilimnion and a higher one in the hypolimnion. At a depth of 1–3 m, we could notice a certain decrease in the conductivity value in all sampling points. The minimum mean values are 160.23 mg/L on the surface and 360.69 mg/L over 9 m depths.

The trophicity degree of the lakes in the main river basins of Romania was determined by the analysis of the chemical aspects (according to the value of the nutrients) and the evolution of the aquatic cenoses (according to the value of the phytoplankton biomass). The qualifications are given by the following grades: hypertrophic, eutrophic, mesotrophic, oligotrophic and ultraoligotrophic, to which intermediate categories are added (Table 8.1).

The eutrophic lakes are characterized by important primary productivity, resulting in a high nutrient content (Coops et al. 2008; Kopácek et al. 2003; Rogora et al. 2001; Rzychoń and Worsztynowicz 2008). On the other side, the oligotrophic lakes have very reduced nutrient content, resulting in low productivity. The mesotrophic lakes are situated between the two categories, with intermediate productivity. The hypertrophic and ultraoligotrophic lakes are situated at the two extremes.

The geographic location of the lakes, in distinct landform units (mountains, hills, and plains, littoral) creates a different environment for the manifestation of trophicity. Thus, according to the value of the nutrients, the greatest parts of the lakes (wetlands) are situated in the mesotrophic, meso-eutrophic, eutrophic, eutro-hypertrophic and hypertrophic categories (17 of the total 19, meaning 88.49 %). Only in the mesotrophic category, 3 lakes (wetlands) are included (15.78 %). In the two extreme categories 1 lake (wetlands) is ultraoligotrphic (5.26 %) and 1 lake (wetlands) is hypertrophic (5.26 %).

According to the phytoplankton biomass value a large number of lakes are included in the categories ultraoligotrophic, oligotrophic and oligo-mesotrophic (13 of 19, representing 68.42 %). In the mesotrophic category, 4 lakes are included (21.05 %). Six lakes are included in the ultraoligotrophic category (31.57.5 %) and 1 lake in the hypertrophic category (5.26 %).

The most hypertrophic lakes are situated in low landform units, where temperatures are high and quantity of nutrients is extremely high (Maneciu). The most favorable area for high trophicity is represented by the Danube flood plain

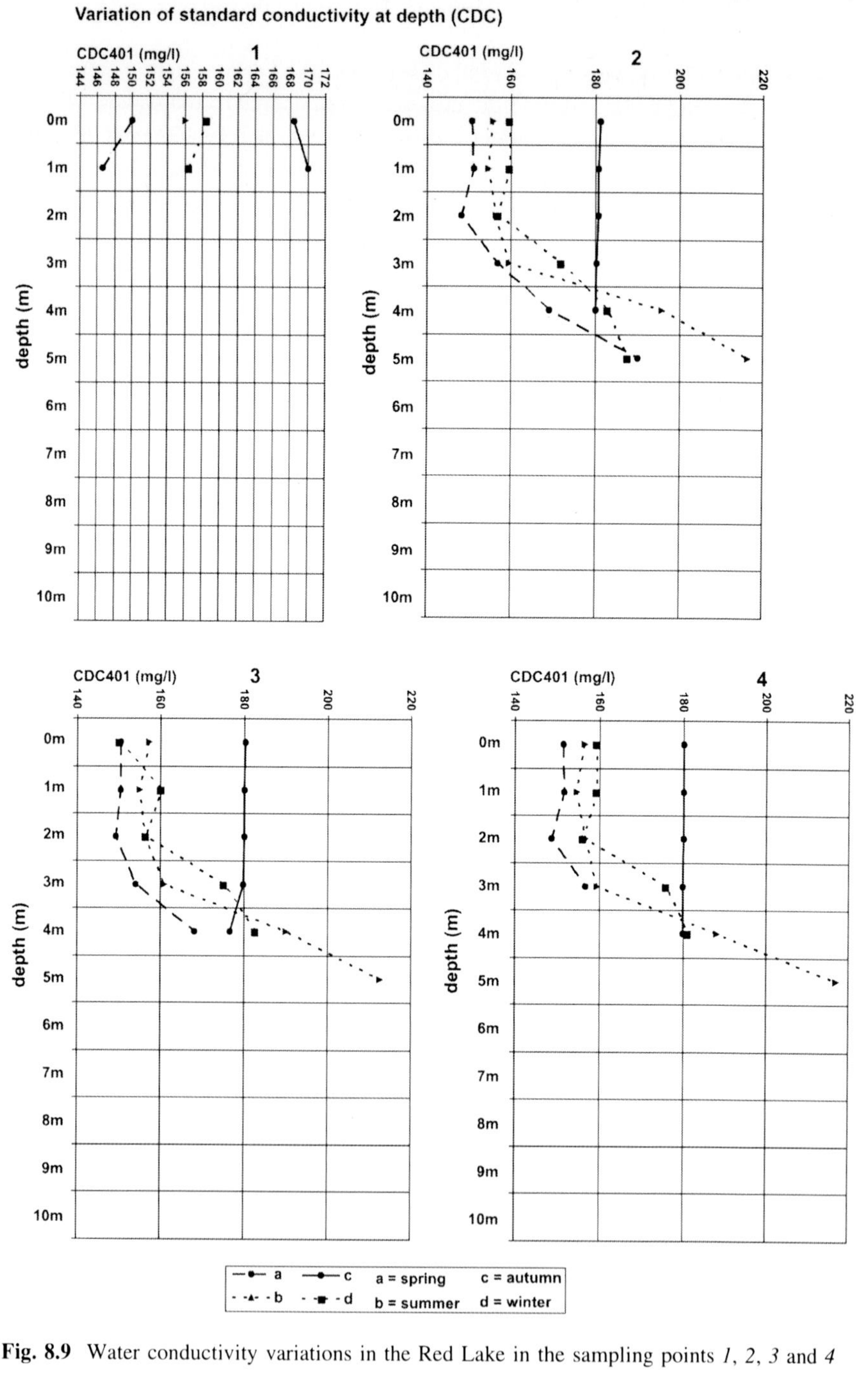

Fig. 8.9 Water conductivity variations in the Red Lake in the sampling points *1*, *2*, *3* and *4*

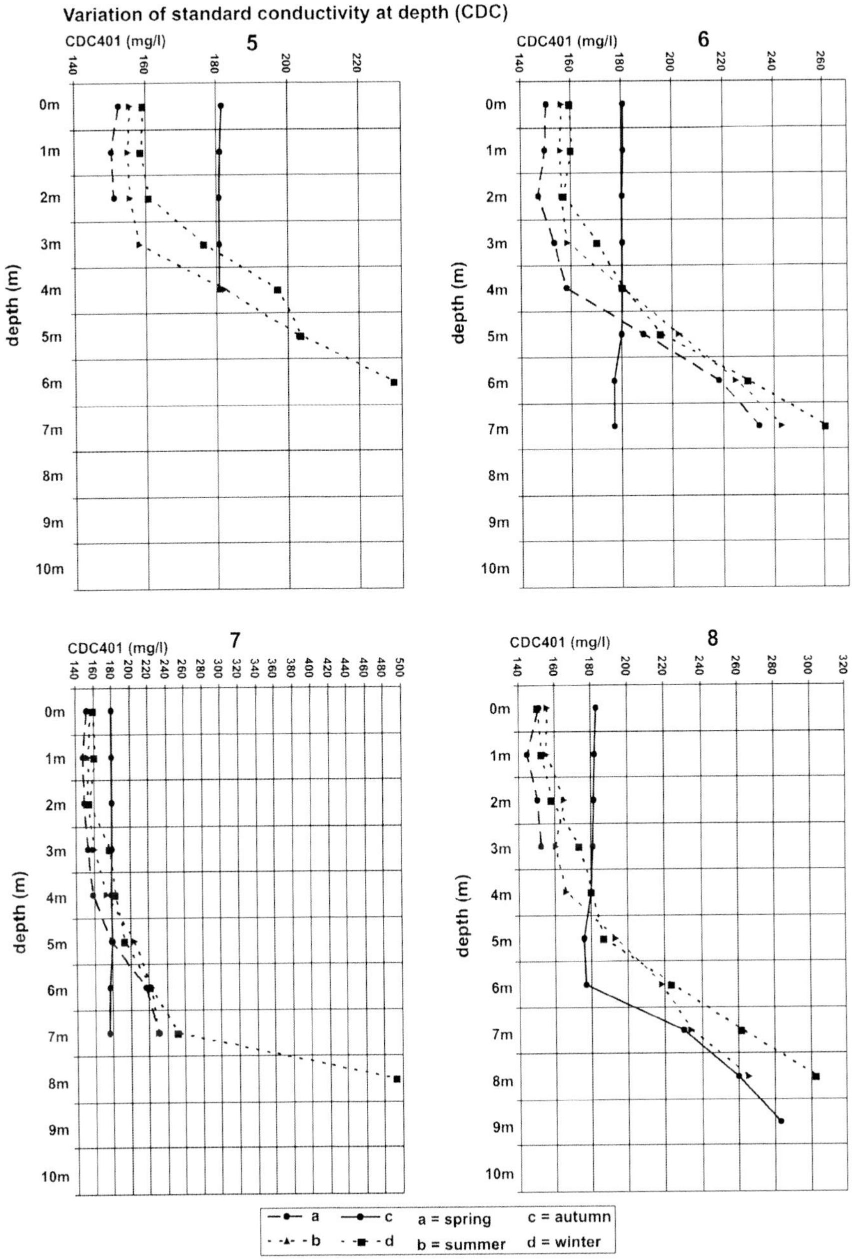

Fig. 8.10 Water conductivity variations in the Red Lake in the sampling points *5*, *6*, *7* and *8*

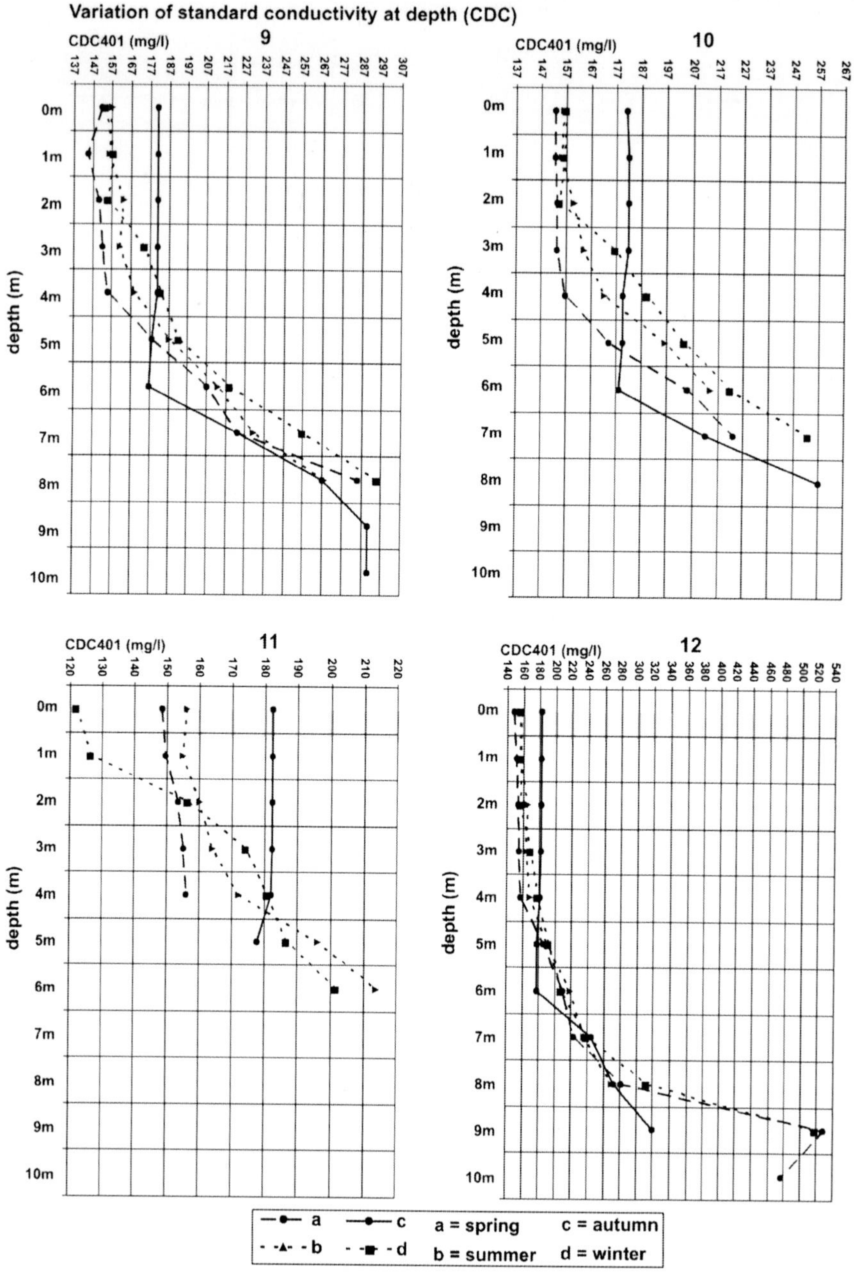

Fig. 8.11 Water conductivity variations in the Red Lake in the sampling points *9*, *10*, *11* and *12*

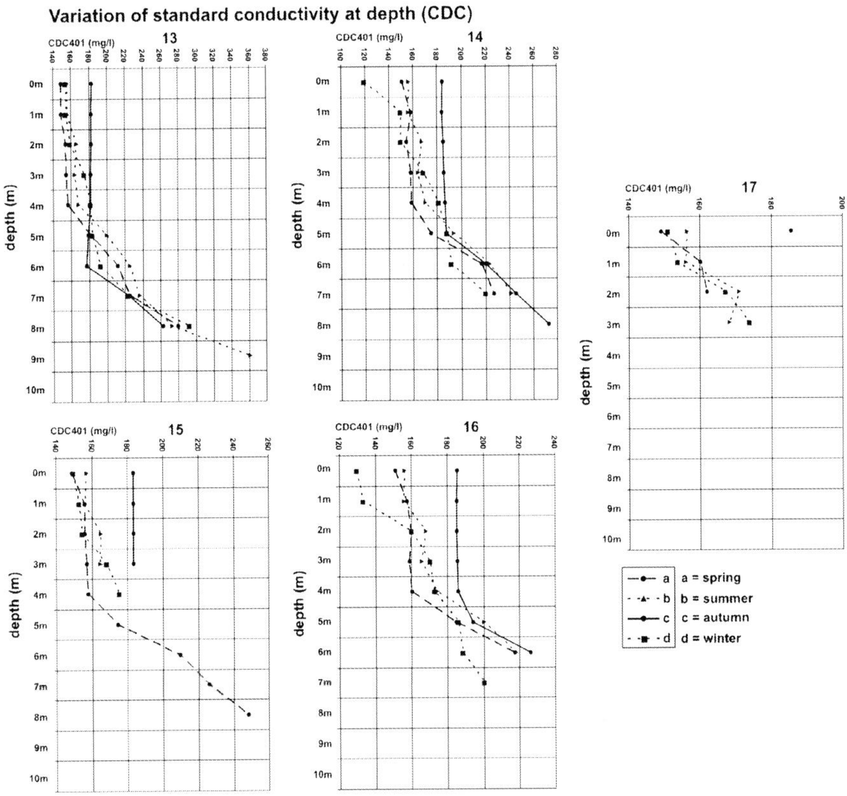

Fig. 8.12 Water conductivity variations in the Red Lake in the sampling points *13*, *14*, *15*, *16* and *17*

(Romanian Academy 1967). The most ultra-oligotrophic lakes are found in the high mountain area. All the glacial lakes are included in this category (Lala). To these, the reservoirs with cold water, developed on hard rocks can be added. In the area with the highest lake density there are lakes (water bodies) with good and very good trophicity (see the Siret and Olt river basins). Because of this reason, many of the accumulation lakes in the Siret and Olt basins are the most important water bodies for the fishing industry.

Besides the climate, trophicity is strongly influenced by the nature of rocks within the drained river basin, erosion rate, transparency and turbidity, etc.

Table 8.1 Morphometric characteristics and trophicity of the lakes in Eastern Carpathians

Lake	Genetic type	Water stream	Usage	Volume mln.m^3	Surface ha	Maximum depth	Water quality (category)	
							Nutrients (total nitrogen, total phosphorous)	Biology
Tisa river basin								
Buhaescu	Glacial	–	–	0.004	0.2	5	M-E	O
Calinesti-Oas	Human made	Tur	Complex	29.0	160	9	E	E
Somes river basin								
Colibita	Human made	Bistrita	Complex	101.2	314	50	E–H	O
Firiza	Human made	Sasar	Complex	–	110.0	37.5	E	O
Lala	Glacial	–	–	–	0.6	2	UO	UO
Strâmtori	Human made	Firiza	Complex	16.6	113.0		E	M
Bodi-Mogosa	Flood plain	Sasar	Complex	0.3	1.6	7	E–H	UO
Olt river basin								
Mesteacanu	Human made	Olt	Water supply	1.1	15	16	M-E	UO
Sfânta Ana	Volcanic	–	Tourism	0.250	19.5	7.5	M	M
Frumoasa	Human made	Frumoasa	Water supply	10.6			O-M	UO
Sacele	Human made	Târlung	Water supply	18.3	–	37	M-E	UO
Bradisor	Human made	Lotru	Complex	38.0	–	–	M	H
Ialomita river basin								
Paltinu	Human made	Doftana	Water supply	62.3	215	107	E–H	O
Maneciu	Human made	Teleajen	Complex	58.0	192	–	H	O
Siret river basin								
Izvorul Muntelui	Human made	Bistrita	Complex	1,230.0	3,000	90	M	M
Bâtca Doamnei	Human made	Bistrita	Energy	10.0	235	16	E	O
Poiana Uzului	Human made	Uz	Water supply	90.0	335	75	E–H	M
Siriu	Human made	Buzau	Complex	158.0	360	–	E	O
Red (Rosu)	Natural barrage	Bicajel	Tourism		13	7.5	M-E	UO

H-hypertrophic; E-eutrophic; M-mesotrophic; O-oligotrophic; UO-ultraoligotrophic

References

Coops H, Buijse L, Buijse ADT, Constantinescu A, Covaliov S, Hanganu J, Ibelings BW, Menting F, Navodaru I, Oosterberg W, Staras M, Torok L (2008) Trophic gradients in a large-river Delta: ecological structure determined by connectivity gradients in the Danube Delta (Romania). River Res Appl 24(5):698–709

Ghenciu V (1972) Trăsături hidrofizice, hidrochimice şi hidrobiologice ale Lacului Roşu. PhD thesis, Faculty of Biology, Iaşi

Kopácek J, Cosby BJ, Majer V, Stuchlík E, Veselý J (2003) Modelling reversibility of central European mountain lakes from acidification: Part II—the Tatra Mountains. Hydrol Earth Syst Sci 7(4):510–524

Rogora M, Marchetto A, Mosello R (2001) Trends in the chemistry of atmospheric deposition and surface waters in the Lake Maggiore catchment. Hydrol Earth Syst Sci 5(3):379–390

Romanian Academy (1967) Limnologie. Sectorul romanesc al Dunarii. Studiu monografic. Academia R.S. Romania, Bucuresti

Rzychoń D, Worsztynowicz A (2008) What affects the nitrogen retention in Tatra Mountains lakes' catchments in Poland? Hydrol Earth Syst Sci 12:415–424

Sobek S, Durisch-Kaiser E, Zurbrugg R, Wongfun N, Wessels M, Pasche N, Wehrli B (2009) Organic carbon burial efficiency in lake sediments controlled by oxygen exposure time and sediment source. Limnol Oceanogr 54(6):2243–2254

Chapter 9
Vegetation and Fauna

Abstract The chapter concerning the vegetation and fauna of the Red Lake (except for the wetlands) is based upon data taken from the scientific literature. From this perspective, the ecologic succession of the lacustrine system may be divided into three phases: open water with an oligotrophic lake regime; invasion of the lake with aquatic macrophytes, with an oligotrophic-mesotrophic lake regime; and, in the end, mesotrophic; senescence, with a mesotrophic-eutrophic lake regime (Ghenciu Trăsături hidrofizice, hidrochimice şi hidrobiologice ale Lacului Roşu, 1972). The vegetation within the hydrographic basin is dominated by pine forests: spruce (*Picea excelsa*); silver fir (*Abies alba*); white pine (*Pinus silvestri*); larch (*Larix decidua*), etc. On the mountain pastures and on the alpine empty crests, graminaceous plants are predominant: red fescue (*Festuca rubra*); matgrass (*Nardus stricta*); rosy vanilla orchid (*Nigritella rubra*) etc., while hydrophyte and helophyte vegetation is present through the following associations: *Carocetum inflatovesicariae; Caricetum appropinquatae; Equisetum limosi* etc. The fauna of the hydrographic basin comprises a great variety of species, many of them rare and protected. Among the mammals the following should be mentioned: the brown bear (*Ursus arctos*); the common wild boar (*Sus scrofa*); the red Carpathian deer (*Cervus elaphus carpaticus*); the roe deer (*Capreolus capreolus*); the wolf (*Canis lupus*); the lynx (*Felis lynx*); the wildcat (*Felis silvestris*); the chamois (*Rupicapra rupicapra*), etc. The fish fauna is specific to the upper salmon area. As it contains dry residue, organic substance, nitrates, phenols, etc., the water provides satisfying conditions for the fish culture.

This chapter concerning the vegetation and fauna of the Red Lake (except for the wetlands) is based upon data taken from the scientific literature. From this perspective, the ecologic succession of the lacustrine system may be divided into three phases: open water with an oligotrophic lake regime; invasion of the lake with aquatic macrophytes, with an oligotrophic-mesotrophic lake regime; and, in the end, mesotrophic; senescence, with a mesotrophic-eutrophic lake regime (Ghenciu 1972) (Fig. 9.1).

G. Romanescu et al., *Limnology of the Red Lake, Romania*,
DOI: 10.1007/978-94-007-6757-7_9,

Fig. 9.1 Compact vegetation with reeds at the outlet of the lake

9.1 The Vegetation

The vegetation within the hydrographic basin is dominated by pine forests: spruce (*Picea excelsa*); silver fir (*Abies alba*); white pine (*Pinus silvestri*); larch (*Larix decidua*), etc. On the slopes with a favorable climate there are mixed forests, with conifers and broad-leaved trees (beech– *Fagus silvatica*; silver birch—*Betula pendula*, sycamore maple—*Acer pseudoplatanus*). In the Bicaz Gorges there are beech woods (Seer and Dezsö 2003).

On the mountain pastures and on the alpine empty crests, graminaceous plants are predominant: red fescue (*Festuca rubra*); matgrass (*Nardus stricta*); Lady's-slipper orchid (*Cypripedium calceolus*); rosy vanilla orchid (*Nigritella rubra*); yellow gentian (*Gentiana lutea*), etc. On the limestones, the following plants were observed: liverleaf (*Hepatica transilvanica*); mountain everlasting (*Antennaria dioica*); edelweiss (*Leontopodium alpinum*), etc.

The fauna of the hydrographic basin comprises a great variety of species, many of them rare and protected.

Among the mammals the following should be mentioned: the brown bear (*Ursus arctos*); the common wild boar (*Sus scrofa*); the red Carpathian deer (*Cervus elaphus carpaticus*); the roe deer (*Capreolus capreolus*); the wolf (*Canis lupus*); the lynx (*Felis lynx*); the wildcat (*Felis silvestris*); the chamois (*Rupicapra rupicapra*), etc.

Avifauna is represented by the following species: the blackbird (*Turdus merula*); the sparrow hawk (*Accipiter nisus*); the wood grouse (*Tetrao urogallus*), etc. The natural character and the degree of reduced pollution ensure the Red Lake is in a good ecologic state. This is the reason why the wild duck acclimated well here (Fig. 9.2). The vegetation comprises the following: aquatic macrophytes; phytoplankton; periphyton. The waters of the Red Lake belong to the oligosaprobic-beta mesosaprobic category, with obvious eutrophication processes.

Hydrophyte and helophyte vegetation is present through the following associations: *Carocetum inflatovesicariae; Caricetum appropinquatae; Equisetum limosi; Equisetum varigati; Deschamsietum caespitose; Scirpetum silvatici; Potametum natantis;* etc. (Ghenciu 1972; Nechita 2003) (Fig. 9.3).

The following species of aquatic macrophytes were found: *Carex rostrata; Carex pseudocyperus; Carex appropinquata; Equisetum limosum; Equisetum palustre; Fontinalis autipyretica; Potamogeton natans; Lemna minor; Menta aquatica; Salix cinerea; Alnus incana; Myriophillum spicatu*; etc. (Ghenciu 1972; Nechita 2003) (Figs. 9.4, 9.5, 9.6).

Fig. 9.2 Wild ducks on the surface of the Red Lake

Fig. 9.3 *Bioskene* on the fossil trunks within the Red Lake

Fig. 9.4 Wetland with alders in the mouth sector of the Oaia (Oii) brook

Fifty nine taxons and 77 species of phytoplankton have been determined. Diatoms are the first, with 51 taxons, followed by the chlorophyceae, with 17 taxons. The following groups have been identified: *Euglenophyta; Cyanophyta; Chrysophyta; Pyrophyta; Chlorophyta; Conjugatophyta; and Bacillariophyta.*

The phytoplankton oscillates depending on the thermal regime: reversed thermal stratification 505 cellules/ml; spring homeothermy 1,051 cellules/ml; direct thermal stratification 1,265 cellules/ml; autumn circulation 922 cellules/ml. The highest values have been registered at the end of summer and the beginning of autumn, as well as at the end of spring (Ghenciu 1972).

Fig. 9.5 The floodplain vegetation of the Oaia (Oii) brook at the lake entrance

Fig. 9.6 *Juncus* vegetation at the Oaia (Oii) brook outlet

There are six unique species in the Romanian algoflora: *Chrysolykos planktonicum; Dinobrion acuminatum; Dinobrion hadzii; Kephyrion rubriclaustri; Kephyrion translucens; Kephyrion spirale.*

Periphyton has the following groups present: *Euglenophyta; Cyanophyta; Pyrophyta; Chrysophyta; Conjugatophyta; Chlorophyta; and Bacillariophyta.* Among the taxons, the following should be mentioned: diatoms 62 %;

Cyanophyceae 15 %; Chlorophyceae 10 %; conjugated 6.3 %; and the rest of the groups 2 % each.

9.2 Fauna

The fauna comprises zooplankton, zoobenthos, fish and batrachians. The zooplankton comprise rotifers, water fleas and copepods. The rotifers contain a great number of species belonging to the following families: *Brachinonidae; Notommatidae; Synchaetidae; Trichoceridae; Lecanidae; Flosculariidae; and Testutinellidae.* The species are cosmopolite, eurybiont, specific to mesotrophic and eutrophic lakes.

Water fleas have species from the following families: *Chydoridae; Bosminidae;* and *Daphniidae.* The number of water fleas varies between 0 and 320,000 individuals/m^3. The lowest number of water fleas is registered in the winter and at the beginning of spring. The maximum number is registered in the months of July, August and September.

The copepods comprise 8 species within the following families: *Cyclopidae* and *Canthocamptinae.* The pelagic plankton organisms are euritherm and eurihaline. The copepods belong to the Paleoarctic. The maximum density is typical to the direct thermal stratigraphy of the months of July and August (Ghenciu 1972).

The benthonic fauna must adapt to the aggradation conditions on the lake bottom. They develop mainly in the areas with macrophytes and on the two main streams (Oaia and Suhard), being better developed in the stations with *Potamogeton* and *Equisetum.* Oligochaeta, chironomidae, nematodes and molluscs are well represented. They reach their maximum number in the summer and at the end of autumn. The average number of zoobenthonic organisms in the stations with macrophytes reaches up to 875–3,070 individuals/m^2. The chironomidae have reached a peak of 7,000 individuals/m^2 (Ghenciu 1972).

The following are specific: Oligochaeta (*Tubifex tubifex, Tubifex costatum, Pristina longiseta*); hirudinea (*Herpobdella oculata, Hermiclepsis marginatus*); bivalve (*Pisidiumcasertarum, Acroloxus lakeustris*); gastropods (*Radix peregra, Radix ovata, Succineea elegans, Mastus venerabilis, Mastus transilvanicus*); ostracods (*Cypris*); decapods; trichoptera; dipters; odonata, etc.

The fish fauna is specific to the upper salmon area. As it contains dry residue, organic substance, nitrates, phenols, etc., the water provides satisfying conditions for the fish culture. The favorable conditions are also given by the abundance of periphyton, macrophytes, zooplankton, zoobenthos and entomofauna. Among these, the salmonidae (*Salmo trutta fario, Salmo trutta lakeustris, Salmo gaidneri* ssp. *irideus*) and the cyprinids (*Phoxinus phoxinus, Rutilus rutilus* ssp. *carpathorossicus*) were observed. The amphibians are represented by *Triturus montandoni, Triturus alpestris, Triturus cristatus, Rana temporaria temporaria, Bufo bufo, Bombina variegate,* etc.

References

Ghenciu V (1972) Trăsături hidrofizice, hidrochimice şi hidrobiologice ale Lacului Roşu. PhD thesis, Faculty of Biology, Iaşi

Nechita N (2003) Flora si vegetatia cormofitelor din masivul Hasmas, Cheile Bicazului si Lacului Rosu (The flora and vegetation of the cormophytes from Hasmas Massif, Bicaz Gorges, and the Red Lake). Bibliotheca Historiae Naturalis II, Natural Sciences Museum, Piatra Neamt, pp 1–383

Seer M, Dezsö L (2003) Lacul Roşu şi Cheile Bicazului. Ghid turistic ilustrat. Mark House, pp 1–50

Chapter 10
The Red Lake Wetland: Boundary Principles and Ecologic Characteristics

Abstract Systemic analysis—mainly the analysis of interactions between elements of a system and the study of the changes between the systems—has become important for current scientific research, known as global or holistic research. Analysis of the "earth-water" system functioning shows the fact that transfers are regulated through transition areas between terrestrial and aquatic environments, known as wetlands (Romanescu et al. 2009, 2010). The wetlands and deep waters of the mountain units within the Eastern Carpathians have suffered fewer transformations compared to those situated in the lower sectors of the Romanian landform. At the level of the entire country, the morphometric characteristics and the climatic conditions allowed the installment of a large variety of wetlands and their conservation has been sustained by conditions preventing drainages or reduced inhabitation density. The most important argument in conserving and rehabilitating wetlands is represented by the roles they have been playing: regulating the hydrologic cycle; increasing the biodiversity; reducing soil erosion; improving the microclimate; reducing the floods; ensuring natural life conditions for numerous species of plants and animals; tourist destination, etc. (Hurt and Carlisle 2001).

Terrestrial, aquatic and aerial compartments constitute a global system and interventions upon one of the elements have repercussions in the functioning of the others. Systemic analysis—mainly the analysis of interactions between elements of a system and the study of the changes between the systems—has become important for current scientific research, known as global or holistic research. On these grounds, other management politics rise, more capable of understanding conflicts and, hence, of solving them in a more promising way, in terms of the sustainable development of territories. The global vision presents the advantage of underlying the role of interfaces between compartments of systems or between the systems themselves—this role has often been neglected as a consequence of theme-specific division. In this case, the water issue is symptomatic: is it possible to monitor the functioning of an aquatic system and to direct the water resources without intervening in soil processing, meaning in the management of the

G. Romanescu et al., *Limnology of the Red Lake, Romania*,
DOI: 10.1007/978-94-007-6757-7_10,

hydrographic basin? Analysis of the "earth-water" system functioning shows the fact that transfers are regulated through transition areas between terrestrial and aquatic environments, known as wetlands (Romanescu et al. 2009, 2010).

In most of the cases, pools, mires, mire lands and peat bogs are associated with a hostile environment. There may be multiple causes for this interpretation. Stagnant waters and unstable soils are, most often, dangerous places. The presence of spontaneous fires in the middle of the night, caused by gas emissions, represents the origin of a death-related mythology, invoked, most of the times, by witches.

The heavy smell of mires is an important factor contributing to their bad reputation. The barely breathable air carries dangerous diseases, malaria being the most important. In the stream of history, the opinions on wetlands have been very diverse. Only at the end of the 19th century did people realize that air was not responsible for carrying malaria but mosquitoes. At the same time, paludian fever caused disasters in many global regions, being still present in very isolated places (Amazonia, Zaire, etc.).

In the 19th century, a genuine conflict began between city inhabitants and farmers. Water circulation within cities, on natural or artificial ways, had already become an obsession. The fear of stagnant waters was collective. On the contrary, the small wetland owners wanted to keep them as the productivity of forages was much bigger than that of other lands. In such cases, farmers were the losers most of the times.

The cleaning-related arguments, mostly agro-economic ones, were primordial for eliminating the wetlands, as they were considered unproductive. The disadvantages were represented by the lack of accessibility, usage limitations (induced by submersion that could last up to a third of a year), the low appetence for certain plants, etc. The dominant plants are cane, reeds, bulrush, etc.

In The Middle Ages, the monks were given several mires in Western Europe. The first large-scale alterations were made for salt exploitation or land tillage (Verger 1968). The monks inside the continents have been, to the same extent, pioneers concerning the alteration of wetlands by creating ponds and stews. Such works had been executed on the rivers close to monasteries. A big part of the Moldavian ponds bears the mark of these transformations (Bendjoudi and Fustec 1996).

The most important drainages took place during the 17th century. King Henry IV promulgated, in 1599, the first administrative act for the drainage of mires in France. The most skilled drainage masters were the Dutch (they practiced poldering). Louis XIII (1613) and Louis XIV (1643), who decided to lease the drained lands, confirmed the ordinances of Henry IV.

In the second half of the 18th century, the wetlands near the banks began to be dammed and drained again. In 1853, Nadault de Buffon proposed a classification "of unsanitary mires." Until World War II, the capitalist society underwent many poldering works along numerous banks, mostly in Western Europe, the USA and South America (Carter 1996).

The same phenomenon was also impelled on the territory of Romania, mostly in the Timiş Plain, where, starting in the 17th century, drainage works were executed on around 850,000 ha, creating a 11,000 km long channel network. The flood-preventing action determined the building of dams with a total length of 1,800 km, the execution of retention basins with a surface of 225,000 ha, numerous stews, etc.

At the end of the 1960s, the drainage and poldering trend inversed and a new trend of "improvement" by drainage for tillage purposes began. Following the 1970 catastrophic floods on the Danube, a vast project of damming the floodplain of the big river began, as well as drainage works for most of the aquatic surfaces. This way, the largest wetlands in Romania were eliminated and the lands were given agrarian purposes. The project itself was based on studies made by authors such as Gr. Antipa (1913, 1921) and Ion Ionescu-Siseşti (Romanescu 2003).

The most diabolic drainage and damming plan was applied during the Communist Era. It targeted the western sector of the Danube Floodplain and it led to the disappearance of 200 lakes (70 in the enclosure of Sireasa and 130 in Pardina). Besides the two floodplain depressions, the Babina and Cernovca enclosures were also drained. Currently, these areas are being proposed for ecologic reconstruction and they have already been re-flooded (Romanescu 1996, Romanescu and Cojocaru 2010).

At a global level, the wetlands occupy around 5 %. The wetlands, at the level of the entire planet, have been profoundly altered, in different ways, in both time and space. Paradoxically, the biological wealth of the wetlands may be compromised through non-intervention. In consequence, the farmers often abandon such wetlands, mostly if there are no access roads or if it costs too much to make it effective in relation to pedologic and hydraulic limitations. In this case, the system closes down and disappears. The abandonment may lead to biological depletion, just like giving the land for culture and intensifying agrarian exploitation (Devillers et al. 1991; Kentula 1996; Wilen et al. 1996).

The wetlands and deep waters of the mountain units within the Eastern Carpathians have suffered fewer transformations compared to those situated in the lower sectors of the Romanian landform. The drainages and agricultural purposes of large wetlands have only occurred in big depressions, where the soil allowed the tillage of plants necessary to local production. In the upper sectors, some wetlands have disappeared due to natural evolution and hydro-technical works have created additional surfaces (Romanescu and Romanescu 2008).

At the level of the entire country, the morphometric characteristics and the climatic conditions allowed the installment of a large variety of wetlands and their conservation has been sustained by conditions preventing drainages or by reduced inhabitation density. The most important argument in conserving and rehabilitating the wetlands is represented by the roles they have been playing: regulating the hydrologic cycle; increasing the biodiversity; reducing soil erosion; improving the microclimate; reducing the floods; ensuring natural life conditions for numerous species of plants and animals; tourist destination, etc. (Hurt and Carlisle 2001).

The hydro-climatic events of the past few years bring back to discussion, or at least they should, the conservation and reconstitution of the former wetlands within the riverside sectors of the main arteries that cause important damage in case of outflows. The drainage of the wetlands has not been specific only to rich countries. The inventory of the wetlands represents only a temporary phase of the assessment. Natural transformations are very rapid and often radical. In order to underline the evolutionary tendency of wetland environments, the inventories should be repeated at certain time intervals.

Nowadays, the mangroves of Asia and Africa are critically endangered. Sixty-seven percent of those of the Philippines disappeared during 60 years (1920–1980). The riverside mires of the Niger have reduced their surface by 300 km^2 after the building of a dam. Since the first European colonists came to the USA, 54 % of the wetlands have been drained (87 mil. ha). In Iowa only 1 % of the original mires have survived. Human intervention upon the wetlands in Europe was much stronger. England lost 60,000 ha per year (between 1970 and 1980) because of drainage and agrarian exploitation. Starting in the 1950s, in England and Wales, 15 % of the seaside mires have been sacrificed for agriculture and industry. In 1978, 84 % of the high peat bogs in the United Kingdom were destroyed because of forestations, reorientation toward cultures and commercial exploitation. In 1940, Spain had 1 mil. ha of wetlands. Between 1950 and 1970, 50 % of this surface was lost (FICWD 1989; Goode 1974).

The disappearance of the wetlands and of the regression rhythm led to a veritable international crusade for their protection. The ecologist movements are the main actors in this fight. In 1980, the first International Conference on Wetlands (New Delhi, India) took place, being organized by INTELCOL (The International Association for Ecology). It distinguished three important phases in the evolution of the ideas related to wetlands:

- underlying the interactions between the terrestrial and aquatic environments;
- comparing the corresponding ecosystems;
- identifying the consequences of destroying the wetlands, mostly the riverside wetlands belonging to big rivers.

Wetlands designate all the continuity elements connecting the aquatic environment with the terrestrial environment. The multitude of difficulties encountered in the process of defining the wetlands is generated by the fact that they are transitional spaces between land and water. We have considered definitions given by different specialists: hydrologists, botanists, paedologists, etc. (Tiner 1996). At the interference between land and water, there is a humidity gradient, which is often continuous, which makes the delimitation of the wetlands more difficult. Furthermore, some wetlands register flooding/land emergence alternations variable in time and space. The amplitude of the flooding phase varies from one year to another, depending on the climatic conditions. In the end, they occupy a permanent spatial and temporal position of a skirt (Lefor and Kennard 1977; Report EPA 1983; Report Y-87-1 1987; Romanescu et al. 2009; Smith 2001).

The term wetland should not be mistaken with the term of aquatic environment. The habitats with deep waters represent the areas that are permanently flooded, situated in the downstream sector or at the margin of the wetlands. In the category of wetlands and habitats with deep waters, the following systems can be distinguished: riverside; palustrine; lacustrine; deltaic; estuarine; marine. Two types are characteristic for the Red Lake: palustrine (with temporary flooding) and lacustrine (with permanent flooding). In the area of the palustrine mires, soil humidity is permanent, even during summertime. The relatively low thermal values cause a low value of evaporability. The influence of salinity is not perceived in any way (Spencer 1994; Sprecher and Warne 2000).

In the delimitation and typology of wetlands, at least four basic factors should be considered: hydrological, geomorphologic, pedologic and biological factors. It is only when the four factors are present that the extent and type of the wetland can be precisely established. Besides the basic conditions, the water physical–chemical parameters are added (Wentworth et al. 1988; Romanescu et al. 2005, 2008).

The topographic measurements for the Red Lake indicated the following values: area—12.01 hectares (120,134.44 m^2); perimeter 2905.79 m; maximum length—1361.72 m; maximum width—160.24 m; maximum altitude upstream in the south (the outlet of the Oaia brook) 965.199 m; maximum altitude upstream in the north-west (the outlet of the Suhard brook) 966.041 m. The length of the main sector (the outlet of the Oii brook and the river mouth) is 958.89 m. The length of the secondary sector (the outlet of the Suhard brook and the river mouth) is 403.32 m.

The Suhard brook slope is big and the bottom of the valley is deep. The smaller distance indicates a higher difference of level. The Oaia (Oii) brook and the associated lacustrine sector implicitly, have a greater length but less slope. In this case, the sedimentation rate could be higher as the current has a lower velocity and the elimination of the alluvia is slower.

The most reduced depths are registered in the upstream sectors of the two branches, at the outlet of the most important brooks: Oaia (Oii) and Suhard. This fact is a result of the relatively reduced solid transport of the two water streams. The reduced solid discharge is the result of a high degree of forest cover that has been maintained since the formation of the lake. Reduced depths are also registered next to the river mouths of the small torrents coming from the mountain sectors with high slopes (Fig. 10.1).

The greatest depths, of 10.5 m, are registered in the central sector, at the confluence of the two main arteries or behind the landslide wave that closed the basin. The higher-level difference between the Suhard brook and the river mouth, causes the appearance of greater depths on this alignment, as compared to the one situated between the Oaia (Oii) brook and the outlet. The water level oscillations registered in the Red Lake are 20–30 cm.

In a transverse profile, a clear asymmetry of the slopes can be noticed. In the remote sector, towards Ghilcos Mountain, from where the landslide material fell off, the slope becomes milder (C), while on the opposite slope (lithological) the slope is very steep (A). The greatest difference of level is registered in the sector of the river mouth (B), where the greatest depths are registered as well (Figs. 10.2, 10.3). A perfect symmetry is registered in the upstream sectors, where the silting is predominant and the quantity of the landslide material is reduced.

The greatest slopes are found on the western and northern banks, where the rocky slopes have an almost vertical fall in the lacustrine basin (Fig. 10.3). These

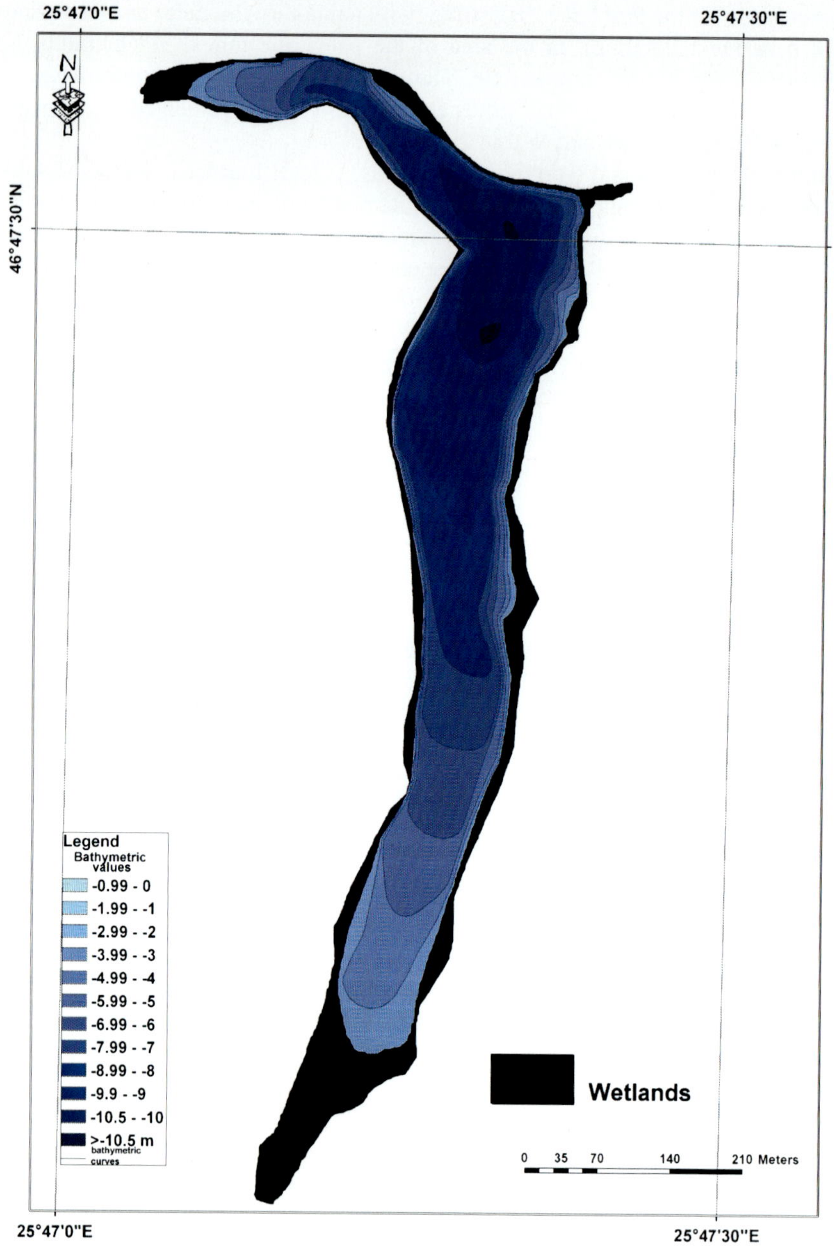

Fig. 10.1 Bathymetric map of the Red Lake and the delimitation of the lacustrine wetland extending to a depth of 3 m (photic layer)

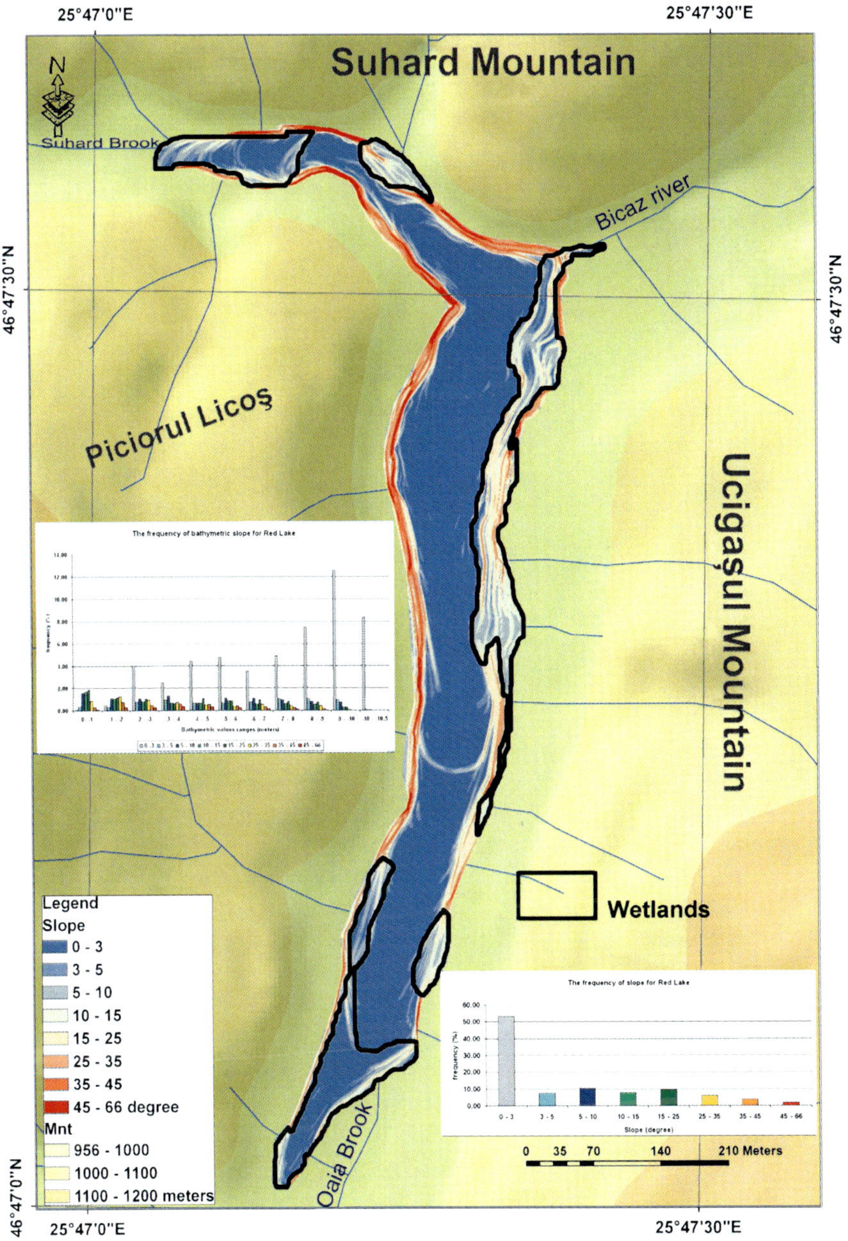

Fig. 10.2 Lacustrine wetlands delimitation according to the value of the submerged slopes

sectors are well forested, they do not represent the source for any significant water stream and the quantity of alluvia material is low. A different situation occurs on the opposite slope, the eastern slope, where the landslide diluvium manifested. In

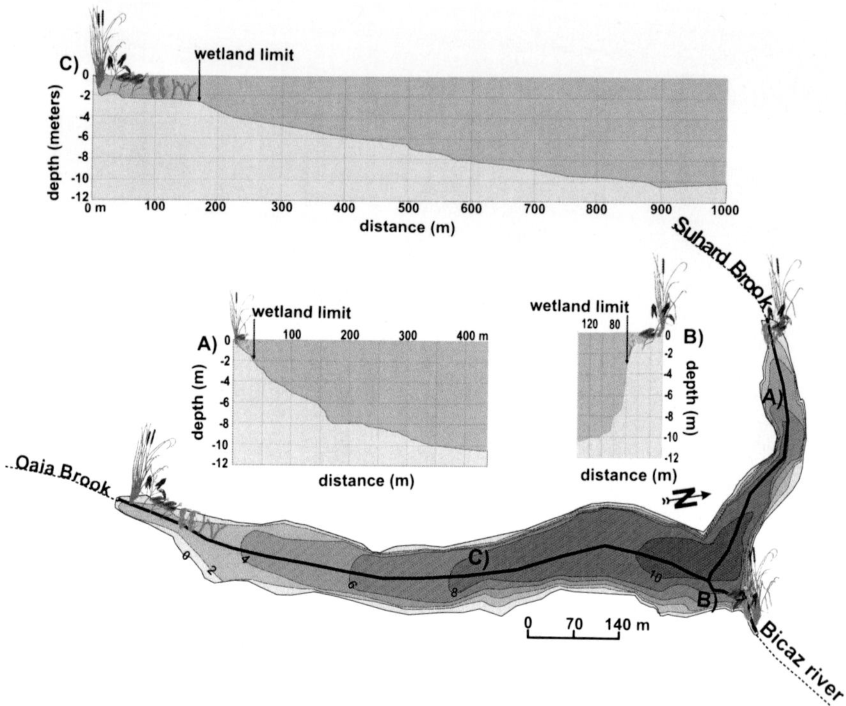

Fig. 10.3 General slope of the Red Lake basin bottom and delineation of the wetlands

this case, the slopes have reduced values and the plant community could develop. This latter sector is strongly fragmented by several torrential arteries. They are anthropically controlled at present, as the linking road crosses this area between Bicaz and Gheorgheni.

The highest percentage is registered by the slopes between 0 and 3° (over 50 %) and the lowest values, by the slopes between 45 and 66° (under 2 %). The slopes between 5–10° and 15–25° register a rather important percentage (almost 10 %). At reduced depths (0–1 m, 1–2 m) the average slopes are predominant, while at high depths (especially for the depths between 8 and 9 m, 9 and 10 m, over 10 m) the reduced slopes are predominant. For the depths exceeding 9 m the slopes between 0 and 3° are predominant, therefore, a uniformity of the bottom lake occurs.

The areas with reduced depths, where light can penetrate up to the bottom of the lake, permitted the installation of vegetation specific to high humidity and therefore, large wetlands occur. Vegetation has its own contribution to the fast siltage of the upstream sectors of the two branches. At the same time, it created a soil that is typical to the lacustrine basins, called limnic soil.

The letter "L" shape is given by the existence of the two important valleys: Oaia (Oii) in the central-southern half and Suhard in the northwestern sector. The landslide body, with a clear asymmetric transverse profile, affects the central-southern branch of the Oaia valley. Landslides do not affect the branch of the

Suhard valley but it has an asymmetric transverse profile due to significant silting towards the northern bank. From this bank, several torrential organisms bring alluvia material and deposit it under the form of small submerged cones of fan-delta type. The same micro-morphological formations can be found on the eastern slope of the main valley as well.

A very important ecologic characteristic for the lacustrine basins is represented by the banks orientation (slope orientation) according to the cardinal points. The orientation of the banks towards the sun or in the shade leads to a different distribution of the lacustrine organisms.

The highest values are characteristic to the banks with eastern, northeastern and northwestward orientation. They follow the general orientation of the lake and its elongated shape. The lowest values are characteristic to the southern and south-western banks.

There are important differences in the orientation of the banks at different depths. The general situation is somehow changed in the middle part of the lake, between the depths of 4–5 m, 5–6 m, 6–7 m, where other orientations are predominant. At higher depths, the eastern and northeastward orientations are predominant and this is due to the inclination of the submerged glacis formed at the base of the lithological slope (hard).

The slope affected by the landslide, with mild slopes, has a general westward orientation, while the opposite bank has an eastward general orientation. The landslide wave blocking the Red Lake basin presents a slope with vertical values, well defined, with little changes along time. Precisely the weak changes permitted the maintenance of the water volume and the existence of the lake itself.

The vertical distribution of the depths creates a transverse profile of the lacustrine basin in the shape of a conical bag. The greatest area values are occupied by isobaths of 0 m. The isobath of −1 m covers an area of 11,236 hectares and the one of −10 m only .056 ha. At the level of each bathymetric curve, an equivalent of solar energy is received. This varies depending on turbidity, transparency, vegetation cover degree, ice sheet cover degree (isolated or continuous), etc.

The general slope of the lake bottom profile is 11° on the alignment of the Oaia brook and the river mouth, 26° between the Suhard brook and the river mouth and 56° on the alignment of the confluence between the two sectors and the main spillway. The greater length can also explain the lower slope in the southern sector. The higher slope of the northern sector is a result of the reduced length. The high slope next to the river mouth is a result of the presence of the extremity of the landslide wave. By analyzing the hypsometric integral for the Red Lake lacustrine basin, one can notice a permanent accumulation and a continuous modification of the submerged slopes because of accretion (bottom and lateral).

Because of the fact that the Red Lake is situated within a mountain area, with relatively steep slopes, the wetlands around it can be easily delimited. Only the eastern slope, affected by the landslide body, induces a small inconvenience. The western (lithological) slope is steep, rocky, with high depths and the wetlands are limited or absent. The same thing is valid for a part of the northern slope. The lack of difficulty on the delimitation and typological framing are characteristic in the two river mouth sectors of the tributary brooks: Oaia (Oii), in the south and Suhard, in the north-west. The same methodology was applied to the river mouths of the small brooks and of the main river mouth.

For delimitation of the sectors included in the category of typical wetlands, the main criteria are used. For geomorphologic delimitation, bathymetric and slope maps should be used. The delimitation of the emersed wetland (palustrine), or submerged wetland, with reduced depths, is simple. The submerged wetland could go as deep as 2–3 m, or, in the present case, it is the photic layer, with a productive bottom (3 m deep). The maximum transparency of the Red Lake is 3 m. The minimum transparency is produced during torrential rains, when the brooks transport a big quantity of alluvia. In this case, the transparency goes down to 10–15 cm. The maximum transparency is registered in the central sector, where the greatest depths are present. From this point of view, three important areas with wetlands of ecologic importance are delimited: the mouth of the Oaia (Oii) brook; the mouth of the Suhard brook; the main river mouth. The wetlands accompanying almost continuously the banks of the lake (except the lithological slopes that fall directly into the lake) have variable size and they are narrow most of the times.

Delimitation according to the slope is relatively similar to delimitation according to bathymetry. Besides the similarity with bathymetry delimitation, one can also notice a greater extension of the wetland associated with the landslide body, with less steep slopes. In the case of the western or northern lithological slopes, the wetlands are limited or absent. The silt samples from the bottom of the lake were taken in 17 points covering the whole surface of the lake. These points are representative for the erosive processes at the surface, as well as for the alluvial processes within the lake basin.

As the water flowing speed reduces, the brooks tributary to the Red Lake gradually deposits alluvia: first, the rolled and dragged ones and then the suspension ones. The former are deposited mainly next to the source and the latter are deposited further away. The lacustrine sediments in the area of the Oaia (Oii) brook spillway have a grey color, with a grey-blackish color towards the surface. The darker color at the surface is a result of higher humus content. Next to the Suhard brook an abundance of craggy elements is registered. In this case, the slope of the river is greater as well. The sediments in the Red Lake present characteristics of fine micaceous sand.

The thickness of the lacustrine sediments varies extremely, reaching a maximum of over 6 m in the spillway sectors of the two important brooks: Oaia (Oii) and Suhard. Typical soils for the wetlands are hydric soils for the emersed sectors and limnic soils for the submerged sectors. For the delimitation of the submerged wetlands, the total organic carbon content and humus content are considered. A clear connection can be noticed between the total organic carbon content and the humus content. At the same time, the connection is natural for the extension of wetlands. At reduced depths, higher organic carbon and humus amounts are present. At greater depths, the finest alluvia are found and they have the most reduced organic carbon and humus quantity.

The more reduced the slopes and depths, the higher the concentration of organic carbon and humus. In the deepest sectors, especially towards the river mouth, smaller quantities are found as well. This fact is due to the great distance to the alluvia source and to the existence of a weak aquatic current, which carries a reduced amount of alluvia. The alluvia are stopped at the two extremities of the lake by the existence of immerse and submerged vegetation, extremely well developed in the torrential rain season (Fig. 10.4).

The greatest value of organic carbon is 7.31 % at the spillway of the Oaia (Oii) brook and 6.10 % at the spillway of the Suhard brook. The most reduced value (3.73) is registered next to the river mouth, where the greatest depths are registered as well. The high values of humus are caused by the existence of rich vegetation in the sectors with reduced depths. In these areas, the water mass is very lighted and the roots of the plants can penetrate the soft layer. These are the most important wetlands around the lake.

The humus in the sediments of the lake comes from two sources: eroded soil from the slopes; accumulation produced by plant decomposition. This way a typical limnic soil is formed. The only comments that could be made are on the genesis and particle size of the sediments. The greatest values of humus are 12.7 at the spillway of the Oaia (Oii) brook and 10.53 at the spillway of the Suhard brook. The lowest value is 6.43 and it is registered at the river mouth, where the greatest values of depth are also registered.

The existence of a significant amount of humus and typical limnic soil, implicitly, facilitated the development of immerse and submerged vegetation

Fig. 10.4 The wetland (lacustrine and palustrine) at the mouth of the Oaia (Oii) brook

(Barbu et al. 1977; Lupaşcu 1999). This fact is normal for depths up to 2–3 m. For greater depths, light hardly penetrates and therefore, the bottom of the lake lacks such life forms. This is the domain of the bacteria and animals that can resist under such conditions created by oxygenation, pH, salinity, pressure, etc.

On the lithological slopes (western and northern) where the slopes are greater and the depths are greater, the lowest quantities of organic carbon and humus are found. This fact is due to the weak erosion of the slopes (perfect forest cover) and to the lack of brooks (therefore a lack of supplementary alluvia). The physical–chemical properties of waters are not too different from the brooks supplying it. There are reduced chemical differences between the waters of the main tributary brooks. According to the classifications used in Romania (Romanescu et al. 2005) the wetlands corresponding to the Red Lake have similar characteristics to the majority of the lakes in the Carpathians (Table 10.1).

Considering the physical-geographic factors mentioned above, the plants create specific associations on the banks of the lake or in the lake (Dobrescu and Ghenciu 1970, 1971a, b; Nechita 2003; Lupascu 1999).

The phytocenosis of *Scirpetum sylvatici* Association Maloc 35 em. Schwich 44 develops on the alluvial soils around the river mouth of the Oaia (Oii) brook and in front of it. The species have a sub-hydrophilic and mesophile character: 38 % meso-hydrophyte, 38 % mesophyte, 19 % hydrophyte, 5 % xero-mesophyte, 2 % ultra-hydrophyte. From a phyto-geographic point of view 41 % are Euro-Asian, 25 circumpolar, 15 cosmopolite, 9 % European, 5 % Central-European and 2 % Carpathian-Balkan. The following species can be mentioned: *Caltha palustris* ssp. *laeta, Deschampsia caespitosa, Equisetum palustre, Poa palustris, Juncus effusus, Lychnis flos cuculi, lycopus europaeus, Agrostis stolonifera, Poa trivialis, Prunella vulgaris, Stellaria graminea, Trifolium repens, Carex acuta, Chaerophyllum hirsutum, Epilobium reseum, Equisetum fluviatile,* etc.

Deschampsietum caespitosae Association Horvatič 30 is specific to the land with profound soils and high humidity by means of gleization. In this case, the characteristic species occupy 80–90 % of the surface. The humus amount is high. It is specific to the river mouths of the Oaia (Oii) and Suhard brooks and to the larger area around the Red Lake. 20 % of the plants are hydrophyte, 34 % meso-hydrophyte, 30 % mesophyte, 8 % xero-mesophyte, 3 % ultra-hydrophyte and 5 % amphy-tollerant. From a phyto-geographic point of view, 51 % are Euro-Asian, 31 % circumpolar, 13 % cosmopolite and 5 % European. Among these the following species can be mentioned: *Mentha longifolia*, *Ranunculus repens*, *Poa trivialis*, *Glechoma hederacea*, *Anthoxanthum odoratum*, *Briza media*, *Festuca pratensis*, *Holcus lanatus*, *Alopecurus pratensis*, *Dactylis glomerata*, *Filipendula ulmaria*, *Trifolium pratense*, etc.

Typhetum angustifoliae Association (All. 22) Pign 43 accompanies the banks of the lake (immerse and submerse). The dominant species is accompanied by *Typha latifolia* but the elements of the banks phytocenosis can also be present. It develops on a clayey soil and limestones substrate. According to the humidity, 13 % are ultra-hydrophyte, 39 % hydrophyte, 32 % meso-hydrophyte, 13 % mesophyte and 3 % xero-mesophyte. From a phyto-geographic point of view 36 % are circumpolar,

Table 10.1 Typological framing of the Red Lake wetlands

Wetland	System	Subsystem	Class	Subclass	Water regime	Salinity	pH	Soil type	Evolution	Protection level	Perspectives
Red Lake	Lacustrine	Limnetic	Nonconsolidated substrate	Silt	Permanently flooded	.0 Sweet	8.04 Alkaline	Organic	+:	2	1

'+' improved sensitive wetland zone, '2' levels of protection of 25–50 %, '1' serious environmental perspective of stabilization

32 % Euro-Asian, 26 % cosmopolite and 6 % European. The following species are observed: *Typha latifolia*, *Glyceria maxima*, *Solanum dulcamara*, *Galium palustre*, *Poa palustris*, *Scutelleria galericulata*, *Carex vulpina*, *Equisetum fluviatile*, *Lycopus europaeus*, *Mentha aquatica*, *Eleocharis palustris*, *Alisma plantago-aquatica*, *Mentha arvensis*, *Scirpus sylvaticus*, *Caltha palustris* ssp. *laeta*, *Epilobium palustre*, *Juncus effusus*, *Lemna minor*, etc.

Typhetum shuttleworthii Association Soó 27 is ultra-hydrophyte and can be found in the shallow waters of the Red Lake. From the point of view of humidity, 31 % are hydrophyte, 31 % meso-hydrophyte, 21 % mesophyte and 17 % ultra-hydrophyte. According to the phyto-geographic origin, 47 % are circumpolar, 35 % Euro-Asian, 14 % cosmopolite and 4 % Central-European. The following species are observed: *Solanum dulcamara*, *Typha latifolia*, *Typha angustifolia*, *Glyceria maxima*, *Lycopus europaeus*, *Alisma plantago-aquatica*, *Equisetum limosum*, *Mentha arvensis*, *Mentha longifolia*, *Myosotis scorpioides*, *Caltha palustris* ssp. *laeta*, *Scutellaria galericulata*, etc.

Glycerietum plicatae Association (Kulcz. 28) Oberd. 52 is developed in the permanently wet areas of the banks, on Calcareous soils, rich in humus. According to the humidity, 47 % are hydrophyte, 33 % meso-hydrophyte, 10 % ultra-hydrophyte, 7 % amphy-tollerant and 3 % mesophyte. According to the soil reaction, 61 % of the plants are adjusted to great variations, 27 % are neutrophile, 9 % acido-neutrophile and 3 % acidophile. According to their origin, 40 % are circumpolar, 27 % Euro-Asian, 27 % cosmopolite and 6 % European. Among these the following species can be mentioned: *Glyceria fluitans*, *Epilobium parviflorum*, *Veronica beccabunga*, *Epilobium roseum*, *Equisetum fluviatile*, *Galium palustre*, *Mentha aquatica*, *Lycopus europaeus*, *Alisma plantago-aquatica*, *Eleocharis palustris*, *Equisetum palustre*, *Equisetum telmateia*, *Poa pratensis*, *Agrostis stolonifera*, *Caltha palustris* ssp. *laeta*, etc. *Carici flavae – Cratoneuretum filicini* Kovács et Felföldy 58 Association includes fragmentary phytocenoses.

Glyceria fluitans, *Epilobium parviflorum*, *Equisetum fluviatile*, *Galium palustre*, *Mentha aquatica*, *Lycopus europaeus*, *Alisma plantago-aquatica*, *Eleocharis palustris*, *Equisetum palustre*, *Equisetum telmateia*, *Caltha palustris* ssp. *laeta*, etc.

Caricetum vesicariae Association Chouard 24 is installed on peaty soils, submerged most of the year. According to the humidity, 52 % are hydrophyte species, 38 % meso-hydrophyte and 10 % ultra-hydrophyte. From a phyto-geographic point of view, 58 % are circumpolar, 14 % cosmopolite, 14 % Euro-Asian and 14 % European. The dominant species are: *Carex rostrata*, *Mentha aquatica*, *Carex acuta*, *Carex appropinquata*, *Epilobium palustre*, *Myosotis caespitosa*, *Scutellaria galericulata*, *Equisetum palustre*, *Lycopus europaeus*, *Eleocharis palustris*, *Scirpus sylvaticus*, *Carex flava*, *Filipendula ulmaria*, *Deschampsia caespitosa*, *Glyceria plicata*, etc.

Equisetetum fluviatilis Association Soó 27 is an ultra-hydrophilic association, covering the river mouths of the Suhard and Oaia (Oii) brooks and the proper banks of the lake. It is developed in the submerged sectors, to a depth of 50 cm. It can also be considered a sub-association subordinated to the *Scirpeto-Phragmitetum* Association (Nechita 2003). Due to the high percentage of *Equisetum fluviatile* species, it was considered to be a distinct association within the *Phragmitetea* class. According to the humidity, 54 % of the species are meso-hydropyte, 34 % hydrophyte, 9 % ultra-hydrophyte and 3 % mesophyte. The phyto-geographic elements are 48 % circumpolar, 33 % Euro-Asian, 11 % cosmopolite, 5 % European and 3 % Central-European. The following species can be mentioned: *Carex acuta, Carex vesicaria, Poa palustris, Agrostis stolonifera, Scutellaria galericulata, Epilobium palustre, Myosotis caespitosa, Galium uliginosum, Alisma plantago-aquatica, Carex gracilis, Equisetum palustre, Lycopus europaeus, Caltha palustris* ssp. *laeta, Carex pseudocypreus, Carex rostrata, Cirsium rivulare*, etc.

Caricetum rostratae Association Rübel 21 prefers peaty, submerse, clayey, oligo-mesotrophic soils. The species are 47 % hydrophile, 32 % meso-hydrophyte, 13 % ultra-hydrophile, 4 % mesophyte and 4 % amphy-tollerant. From a phytogeographic point of view, 47 % are circumpolar, 36 % Euro-Asian, 13 % cosmopolite and 4 % European. Among these the following species can be mentioned: *Carex hitra, Carex vesicaria, Galium palustre, Carex pseudocyperus, Epilobium palustre, Myosotis caespitosa, Scutellaria galericulata, Galium uliginosum, Equisetum palustre, Veronica anagalis-aquatica, Typha latifolia, Alisma plantago-aquatica, Caltha palustris* ssp. *laeta, Deschampsia caespitosa, Filipendula ulmaria, Mentha aquatica*, etc.

Caricetum appropinquata Association (Koch 26) Soó 28 can be found on the banks of the Red Lake and at the river mouth of the Suhard brook. It prefers swampy soils, acido-neutral, temporarily submersed, with hydromorphic humus. According to the humidity, 50 % of the species are hydrophyte, 42 % meso-hydrophyte, 4 % ultra-hydrophyte and 4 % amphy-tollerant. From a phytogeographic point of view, 55 % of the species are circumpolar, 29 % Euro-Asian, 12 % cosmopolite and 4 % European. The following species were observed: *Carex acuta, Carex hitra, Carex rostrata, Carex elongata, Epilobium palustre, Myosotis scorpioides, Scutellaria galericulata, Eleocharis palustris, Equisetum palustre, Mycopus europaeus, Veronica anagallis aquatica, Ranunculus repens, Deschampsia caespitoasa, Caltha palustris* ssp. *laeta, Epilobium roseum*, etc.

Limnetum minoris Association (Oberd. 57) Müller et Görs. 60 includes the hydrophilic groups that form compact masses but with sporadic distribution at the edges of the water. The submerged layer includes *Myriophyllum spicatum*. The majority of the elements are ultra-hydrophile, circumpolar or cosmopolite and are developed on weakly acido-neutrophile soils. Among these the following species can be mentioned: *Potamogeton natans, Callitriche cophocarpa, Lemna trisulca, Myriophyllum spicatum, Alisma plantago-aquatica, Glyceria plicata, Juncus effusus, Typha latifolia*, etc.

Batrachio trichophyllo-Callitrichetum polymorphae Association Soó 27 is present in the water of the Red Lake, next to the banks and in the water with reduced depth at the river mouths. The majority of the species are ultra-hydrophyte and prefer eutro-mesophyte soils rich in calcium. The following plants are observed: *Calitriche palustris, Potamogeton natans, Myriophyllum spicatum, Lemna minor, Alisma plantago-aquatica, Glyceria plicata, Ranunculus repens*, etc.

Carici flavae—Cratoneuretum filicini Association Kovács et Felföldy 58 includes fragmentary phytocenosis. It colonizes the Calcareous and permanently swampy substrate, under the form of biochars on the fossil trunks of the lake. It is found at the river mouth of the Suhard brook. According to the humidity, 43 % of the plants are meso-hydrophyte, 30 % hydrophyte, 23 % mesophyte, 2 % ultra-hydrophyte and 2 % xerophyte. From a phyto-geographic point of view, 36 % of the plants are circumpolar, 32 % Euro-Asia, 12 % European, 8 % Central-European and 12 % cosmopolite. Among these the following species can be mentioned: *Cratoneurum commutatum, Equisetum telmateia, Pinguicula vulgaris, Viola biflora, Caltha palustris* ssp. *laeta, Cardamine amara, Chrysosplenium alternifolium, Deschampsia caespitosa, Carex serotina, Epilobium palustre, Juncus effusus, Lxcopus europaeus*, etc.

Angelico-Cirsietum oleracei Association Tx. 37 has a fragmentary distribution, under the form of wet or swampy meadows, on the banks of the Red Lake. The soil is hydromorphic, gleyed and wetted by stagnant or infiltration waters. The species include 45 % meso-hydrophyte, 25 % mesophyte, 18 % hydrophyte, 5 % xero-mesophyte and 7 % amphy-tollerant. The structure of the phyto-geographic elements is: 58 % Euro-Asian, 25 % circumpolar, 7 % Central-European, 5 % European and 5 % cosmopolite. The following species can be mentioned: *Chaerophyllum hirsutum, Filipendula ulmaria, Galium palustre, Mentha longifolia, Equisetum palustre, Lychnis flos cuculi, Carex ovalis, Cirsium palustre, Carex hirta, Juncus inflexus, Lotus corniculatus, Ajuga reptans, Prunella vulgaris, Carex acuta, Carex flava, Lycopus europaeus*, etc.

Potametum natantis Association Soó 27 prefers the shadowed places of the banks. In includes two strata: natant (*Potamogeton natans, Lemna minor*) and submerged (*Myriophyllum spicatum*). The majority of the species are ultra-hydrophyte, circumpolar or cosmopolite. Among these the following species can be mentioned: *Lemna minor, Callitriche palustris, Myriophyllum spicatum, Typha schuttleworthii, Equisetum telmateja, Glyceria plicata*, etc.

The trees are identical in all the wetlands: *Populus alba, Alnus glutinosa* and rarely *Salix alba*.

Considering all these conditions and using the aerial images (ortophotoplans) an exact delimitation of the most important emersed and submerged wetlands in the three sectors was done (Fig. 10.5). In this case, only the wetlands with ecologic importance were chosen, as they are weakly modified by human interference. The delimited surfaces cover an area of 4.0330 ha at the mouth of the Oaia (Oii) brook (Fig. 10.6), 0.3067 ha at the river mouth (Fig. 10.7) and 0.1643 ha at the mouth of the Suhard brook (Fig. 10.8). The total area of the wetlands with great ecologic importance is 4.5040 ha.

The beginning of slight deforestation and the building up of some houses and boarding houses have caused significant siltage processes and a slight eutrophication of waters. The case is not very severe but warning should be given in order to preserve clean waters and preserve optimum conditions in the wetlands. Located in the mountainous Eastern Carpathians, the Red Lake falls in the water body category with alpine specifics. In the catchment area, the conical form facilitates

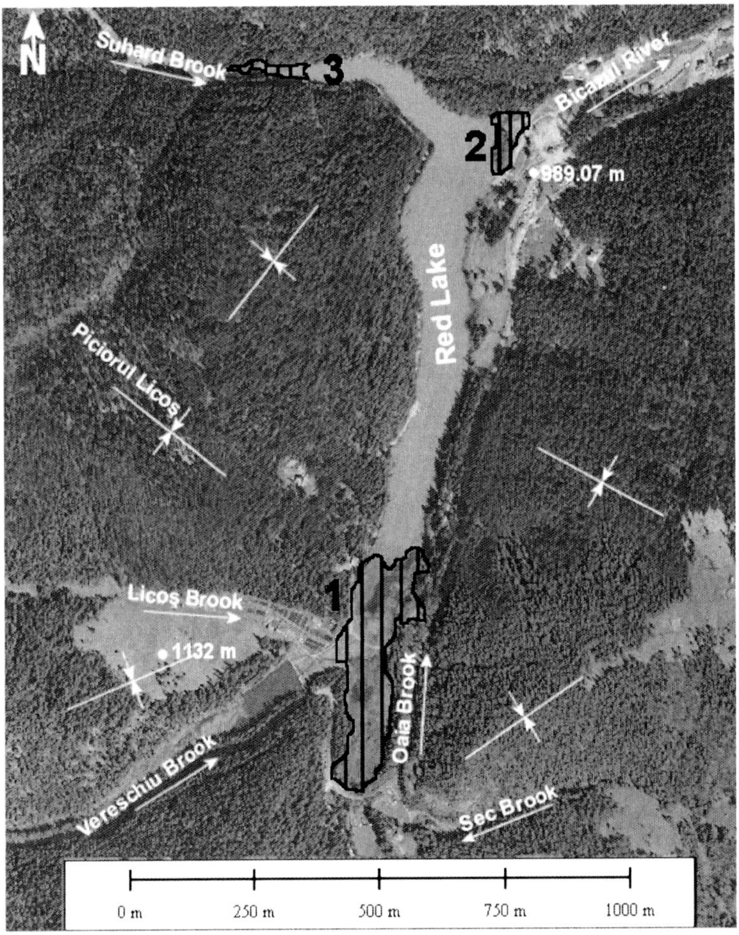

Fig. 10.5 Delimitation of compact wetlands of great ecologic importance (lacustrine and palustrine)

the appearance of thermal inversions with a cold winter climate and another moderately cold climate in summer. It is located at an altitude of 965 m, has a length of 1362.21 m and has a maximum width of 160.24 m.

At least four basic factors, hydrological, geomorphologic, pedologic and biological, were considered in the delimitation of the lacustrine and riparian wetlands. For the wetlands, the bottom of the lake must be productive, that is photic, with the development of water plants. In this case, the light penetrates as deep as 3 m. The wetlands cannot be delimited when one of the four defining factors is missing. The absence of one factor automatically causes an incorrect indexation of that category of wetlands.

Because the surface of the Red Lake river basin is well forested, with conifers in the major part, few hydrographic arteries degrade the surface of the slopes.

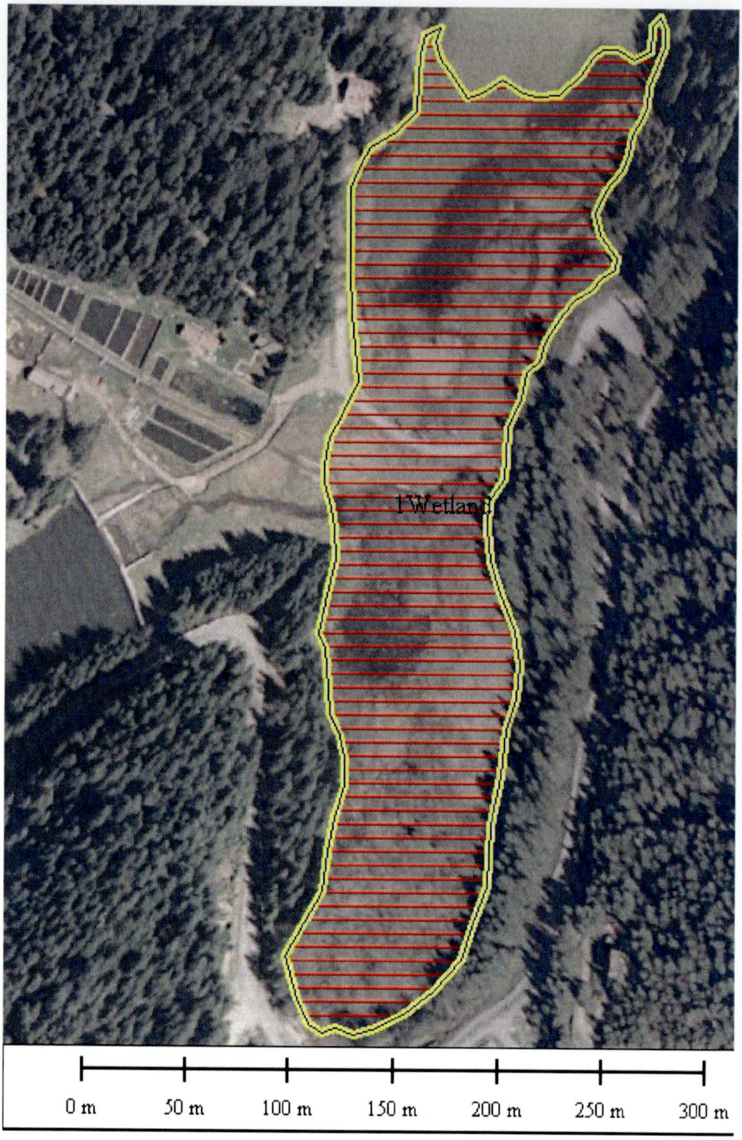

Fig. 10.6 Delimitation of the wetland (lacustrine and palustrine) at the mouth of the Oaia (Oiii) brook

Three important water stream tributaries of the lake basin and one river, a tributary of the river mouth, created the Bicaz River. The waters of the Red Lake or the brooks flood the wetlands analyzed permanently or seasonally. At the same time, they are strongly influenced by the phreatic waters that rise through capillarity from the lake or brooks, or flow down the slopes as a result of gravitation. Thus,

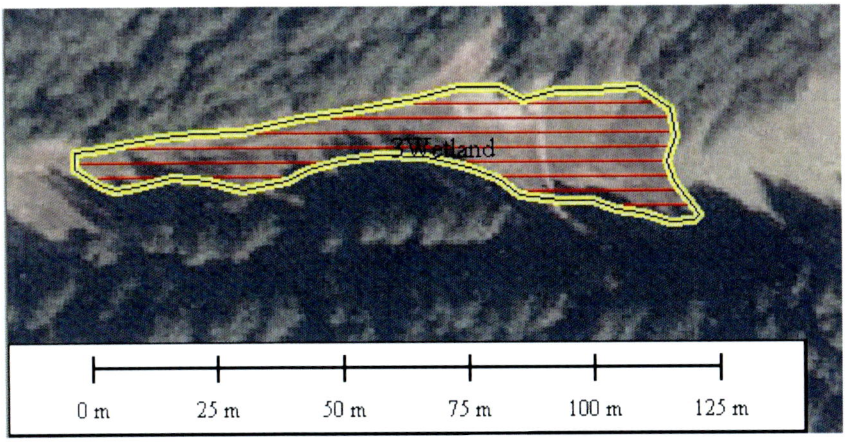

Fig. 10.7 Delimitation of wetlands (lacustrine and palustrine) at the mouth of the Suhard brook

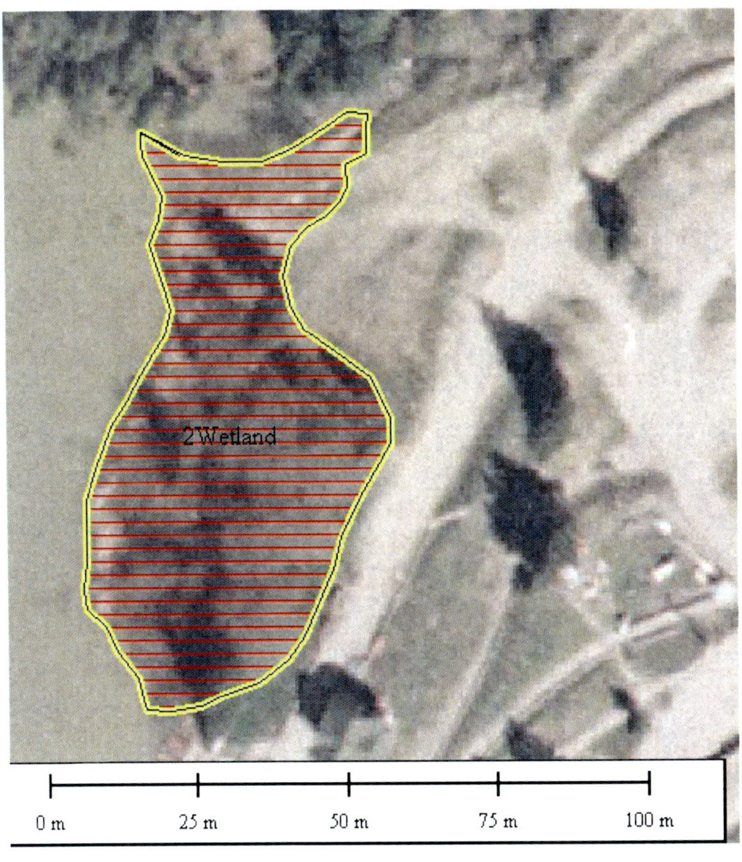

Fig. 10.8 Delimitation of wetland at the river mouth

we cannot consider an increase or decrease of the wetlands, as the delimitation is made strictly based on the average values. Physical–chemical analyses of water were carried out in order to establish the characteristic parameters, which allow the development of certain plant or animal communities. In the present case, there are valid indicators for all plant communities but typical animal communities were not identified for the Red Lake area. The synthetic tables and the extra data reveal a small differentiation between the wetlands in the Red Lake (lacustrine or palustrine) and the typical riparian wetlands (Table 10.2).

Level variations in the Red Lake are low. The rise and fall lie within a range of 20–30 cm. For high waters, during the torrential rains, turbidity is high and the water level can rise up to 30 cm in the upstream sectors. During years of high wind velocity, the water level can fall up to 20 cm. The physical–chemical parameters (TDS—Total Dissolved Solids; U—Tension of Water; LDO—Luminiscent Dissolved Oxygen) in the Red Lake are represented by: TDS—151.6 mg/L; U- −56.8 mV; LDO- 9.23 mg/L (multiannual averages).

The Oaia (Oii) brook makes a direct contribution to the supply of the Red Lake water body with water. As a right tributary of the lake, it has a relatively high slope. With the highest flow of a Red Lake tributary brook, it has a permanent character. Before flowing into the lake, it meets the Licas brook. Together they create a large palustrine wetland. The physical–chemical parameters are represented by TDS—77 mg/L; U- −109.2 mV; LDO—6.8 mg/L.

The Licas brook meets the Oaia (Oii) brook before it flows into the Red Lake. As a left tributary of the lake, it has a permanent character and is the third longest of the tributary brooks. It makes a significant contribution to the maintenance of the hydrologic balance within this water body. There is a trout fishery in its lower stream. The physical–chemical parameters are represented by TDS—63 mg/L; U- −110.7 mV; LDO—8.7 mg/L.

The Suhard brook, relatively well preserved from an ecologic point of view, supplies the Red Lake in the northwestern sector. It creates an elongated wetland on its lower stream. Being the second longest brook, it presents the highest slope of all the tributary brooks. For this reason, the sediments deposited at the river mouth present a coarser fraction as compared to the other hydrographic arteries. The physical–chemical parameters are represented by TDS—96 mg/L; U- −108.3 mV; LDO—5.6 mg/L.

From a geomorphologic point of view, the Red Lake basin stretches over two alignments represented by the most important valleys: Oaia (Oii) and Suhard (Fig. 5.34, p.110). The depths are significant in the central parts (with a maximum of 10.5 m) and reduced next to the river mouths. The depths are more reduced on the bank affected by the landslides, while on the lithological banks (with hard rocks) the depths are greater. The same is valid for the submerged slopes where plant and animal communities are present. The sediments have great thickness in the areas where sedimentation was significant. In these locations, the quantities of total organic carbon were high and facilitated the development of specific vegetation (Fig. 10.9).

Table 10.2 Typological framing of the Red Lake, the Oaia brook, the Licas brook and the Suhard brook wetlands

Wetland	System	Subsystem	Class	Subclass	Water regime	Salinity	pH	Soil type	Evolution[a]	Protection level[b]	Perspectives[c]
Red Lake	Lacustrine	Limonitic	Non-consolidated substrate	Silt	Permanently flooded	.0 Sweet	8.04 Alkaline	Organic	+:	2	1
Oaia (Oii) Brook	Riparian	Perennial lower stream	River bed	Gravel Sand	Permanently flooded	.0 Sweet	9.02 Alkaline	Mineral	=:	3	2
Licas Brook	Riparian	Perennial lower stream	River bed	Gravel Sand	Permanently flooded	.0 Sweet	9.02	Mineral	=:	3	2
Suhard Brook	Riparian	Perennial lower stream	Fluviatile riverbed	Gravel Sand	Permanently flooded	.0 Sweet	8.97 Alkaline	Mineral	++:	2	1

[a] '+' improved sensitive wetland zone, '++' with strong organic improved rich wetlands, '=' a wetland with little alteration

[b] '2' levels of protection of 25–50 %, '3' levels of protection of 15–25 %

[c] '1' serious environmental perspective of stabilization, '2' green areas with an uncertain future

Fig. 10.9 Wetland vegetation at the mouth of the Oaia brook

Finally, the plants form specific associations, close to the banks of the Red Lake or inside the lake, according to the four defining factors (Nechita 2003; Lupascu 1999). The humus amount is high. The following plant associations have been delineated: *Telekio speciosae—Alnetum incanae* Association Coldea (86) 90; *Scirpetum sylvatici* Association Maloc 35 em. Schwich 44; *Festucetum pratensis* Association Soó 38; *Deschampsietum caespitosae* Association Horvatič 30; *Typhetum angustifoliae* Association (All. 22) Pign 43; *Typhetum shuttleworthii* Association Soó 27; *Glycerietum plicatae* Association (Kulcz. 28) Oberd. 52; *Caricetum vesicariae* Association Chouard 24; *Equisetetum fluviatilis* Association Soó 27; *Caricetum rostratae* Association Rübel 21; *Caricetum appropinquata* Association (Koch 26) Soó 28; *Limnetum minoris* Association (Oberd. 57) Müller et Görs. 60; *Batrachio trichophyllo-Callitrichetum polymorphae* Association Soó 27; *Carici flavae—Cratoneuretum filicini* Association Kovács et Felföldy 58; *Angelico-Cirsietum oleracei* Association Tx. 37; *Potametum natantis* Association Soó 27; *Telekio-Petasitetum hybridi* Association (Morariu 67) Resm. et Ratiu 74; *Eleocharidetum palustris* Association Sennikov 19 (Table 10.3).

Depending on weather and morpho-hydrographic conditions and specific plant associations wetlands are landforms distributed in areas of low relief near water surfaces in riverbeds or lake basins. Very absent on steep slopes, they consist of hard rocks: west and north. The larges areas are found at the mouths of two streams, the Oaia and the Suhard (Fig. 10.10).

Plants in the lake basin find favorable conditions in waters with depths of up to 2 m, with an oozy substrate rich in humus. The most important water streams

Table 10.3 The plant associations within the framework of wetlands afferent to the Red Lake

Association	Species	Place	Substratum	Hydrologic reaction	Phyto-geographic origin	Soil reaction
Telekio speciosae—Alnetum incanae Coldea (86) 90	*Salix silesiaca* *Carex remota* *Equisetum Hyemale* *Carex sylvatica* *Mentha longifolia* *Caltha palustris* ssp. *Laeta*	Licas brook Oaia (Oii) brook Bicaz brook	Alluvial soils with high percentages of gravel	50 % mesophyte 32 % meso-hydrophyte 9 % xero-mesophyte 6 % hydrophyte 3 % amphy-tolerant	45 % are Euro-Asian 15 % circumpolar 12 % European 10 % cosmopolite 9 % Central-European 7 % Carpathian-Balkan 2 % Ponto-Mediterranean	45 % weakly acido-neutrophile 31 % no requirements towards this indicator 19 % prefer an acido-neutrophile 3 % are acidophile 2 % neutro-basiphyte
Scirpetum sylvatici Maloc 35 em. Schwich 44	*Caltha palustris* ssp. *Laeta* *Equisetum palustre* *Poa palustris* *Juncus effuses* *Prunella vulgaris* *Trifolium repens* *Carex acuta* *Chaerophyllum hirsutum*	Oaia (Oii) brook: river mouth	Alluvial soils	38 % meso-hydrophyte 38 % mesophyte 19 % hydrophyte 5 % xero-mesophyte 2 % ultra-hydrophyte	63 % euriionic 15 % weakly acido-neutrophile 5 % acido-neutrophile and acidophile 2 % neutro-basiphile	41 % are Euro-Asian 25 circumpolar 15 cosmopolite 9 % European 5 % Central-European 2 % Carpathian-Balkan

(continued)

Table 10.3 (continued)

Association	Species	Place	Substratum	Hydrologic reaction	Phyto-geographic origin	Soil reaction
Festucetum pratensis Soó 38	*Festuca pratensis* *Deschampsia caespitosa* *Mentha longifolia* *Carex pallescens* *Lotus corniculatus* *Equisetum fluviatile* *Caltha palustris* ssp. *Laeta* *Carex echinata*	North-west of the Red Lake	Alluvial, wet, non-evolved soils	43 % mesophyte 30 % meso-hydrophyte 15 % hydrophyte 2 % ultra-hydrophyte 5 % xero-mesophyte 5 % amphy-tollerant	48 % Euro-Asian 30 % circumpolar 10 % cosmopolite 10 % European 2 % Central European	66 % are euriionic 15 % are acido-neutrophile 12 % weakly acido-neutrophile 5 % acidophile 2 % very acidophile
Deschampsietum caespitosae Horvatič 30	*Mentha longifolia* *Ranunculus repens* *Anthoxanthum odoratum* *Festuca pratensis* *Holcus lanatus*	River mouths: Oaia (Oii) brook, Suhard brook. Larger area around the Red Lake	Profound soils and high humidity by means of gleization	20 %hydrophyte 34 % meso-hydrophyte 30 % mesophyte 8 % xero-mesophyte 3 % ultra-hydrophyte 5 % amphy-tolerant	51 % are Euro-Asian 31 % circumpolar 13 % cosmopolite 5 % European	65 % euriionic 17 % weakly acido-neutrophile 15 % acido-neutrophile 3 % acidophile
Typhetum angustifoliae (All. 22) Pign 43	*Typha latifolia* *Solanum dulcamara* *Galium palustre* *Poa palustris* *Carex vulpine* *Equisetum fluviatile* *Mentha aquatic* *Eleocharis palustris* *Alisma plantago aquatic*	Banks of the lake (immersed and submerged)	Clayey soil Calcareous substrate	13 % ultra-hydrophyte 39 % hydrophyte 32 % meso-hydrophyte 13 % mesophyte 3 % xero-mesophyte	36 % circumpolar 32 % Euro-Asian 26 % cosmopolite 6 % European	54 % euriionic 30 % weakly acido-neutrophile 10 % acido-neutrophile 6 % acidophile

(continued)

Table 10.3 (continued)

Association	Species	Place	Substratum	Hydrologic reaction	Phyto-geographic origin	Soil reaction
Typhetum shuttleworthii Soó 27	*Solanum dulcamara* *Typha latifolia* *Typha angustifolia* *Glyceria maxima* *Alisma plantago aquatic* *Equisetum limosum* *Mentha arvensis*	Shallow waters of the Red Lake	Alluvial soils	31 % hydrophyte 31 % meso-hydrophyte 21 % mesophyte 17 % ultra-hydrophyte	47 % circumpolar 35 % Euro-Asian 14 % cosmopolite 4 % Central-European	67 % euriionic 27 % weakly acido-neutrophile 3 % acido-neutrophile 3 % acidophile
Glycerietum plicatae (Kulcz. 28) Oberd. 52	*Glyceria fluitans* *Epilobium parviflorum* *Equisetum fluviatile* *Galium palustre* *Mentha aquatic* *Lycopus europaeus* *Alisma plantago-aquatica* *Eleocharis palustris*	Wet areas of the banks	Calcareous soils that are rich in humus	47 % hydrophyte 33 % meso-hydrophyte 10 % ultra-hydrophyte 7 % amphy-tollerant 3 % mesophyte	40 % circumpolar 27 % Euro-Asian 27 % cosmopolite 6 % European	61 % great variations 27 % neutrophile 9 % acido-neutrophile 3 % acidophile
Caricetum vesicariae Chouard 24	*Carex rostrata* *Mentha aquatic* *Carex acuta* *Carex appropinquata* *Epilobium palustre* *Myosotis caespitosa* *Equisetum palustre*	Shallow waters of the Red Lake	Peaty soils, submerged most of the year	52 % hydrophyte 38 % meso-hydrophyte 10 % ultra-hydrophyte	58 % circumpolar 14 % cosmopolite 14 % Euro-Asian 14 % European	62 % high variations 28 % weakly acido-neutrophile 10 % acidophile

(continued)

Table 10.3 (continued)

Association	Species	Place	Substratum	Hydrologic reaction	Phyto-geographic origin	Soil reaction
Equisetetum fluviatilis Soó 27	*Carex acuta* *Carex vesicaria* *Poa palustris* *Epilobium palustre* *Alisma plantago-aquatica* *Carex gracilis* *Equisetum palustre*	River mouths: Suhard brook, Oaia (Oii) brook Banks proper of the Red Lake	Submerged alluvial soils	54 % meso-hydropyte 34 % hydrophyte 9 % ultra-hydrophyte 3 % mesophyte	48 % circumpolar 33 % Euro-Asian 11 % cosmopolite 5 % European 3 % Central-European	56 % are indifferent 24 % weakly acido-neutrophile 11 % acido-neutrophile 9 % acidophile
Caricetum rostratae Rübel 21	*Carex hitra* *Carex vesicaria* *Galium palustre* *Carex pseudocyperus* *Epilobium palustre* *Equisetum palustre* *Veronica anagalis-aquatica* *Typha latifolia*	Shallow waters of the Red Lake	Peaty submersed Clayey mesotrophic soils Oligo-mesotrophic soils	47 % hydrophile 32 % meso-hydrophyte 13 % ultra-hydrophile 4 % mesophyte 4 % amphy-tolerant	47 % circumpolar 36 % Euro-Asian 13 % cosmopolite 4 % European	57 % euriionic 26 % weakly acido-neutrophile 9 % acidophile 4 % acido-neutrophile 4 % neutro-basiphile
Caricetum appropinquata (Koch 26) Soó 28	*Carex acuta* *Carex hitra* *Carex rostrata* *Carex elongate* *Epilobium palustre* *Eleocharis palustris* *Equisetum palustre*	Banks of the Red Lake River mouth of the Suhard brook	Swampy soils Acido-neutral soils, temporarily submerged, with hydromorphic humus	50 % hydrophyte 42 % meso-hydrophyte 4 % ultra-hydrophyte 4 % amphy-tolerant	55 % circumpolar 29 % Euro-Asian 12 % cosmopolite 4 % European	63 % euriionic 29 % acido-neutrophile 4 % acidophile 4 % neutro-basiphile

(continued)

Table 10.3 (continued)

Association	Species	Place	Substratum	Hydrologic reaction	Phyto-geographic origin	Soil reaction
Limnetum minoris (Oberd. 57) Müller et Görs. 60	*Potamogeton natans* *Lemna trisulca* *Alisma plantago aquatic* *Juncus effuses* *Typha latifolia* *Myriophyllum spicatum*	Sporadic distribution at the edges of the water	–	100 % ultra-hydrophile	50 % circumpolar 50 % cosmopolite	Weakly acido-neutrophile
Batrachio trichophyllo-Callitrichetum polymorphae Soó 27	*Calitriche palustris* *Potamogeton natans* *Myriophyllum spicatum* *Lemna minor* *Alisma plantago-aquatica*	In the water of the Red Lake Next to the banks and in the water with reduced depth at the river mouths	Alluvial soils Limnosoils	100 % ultra-hydrophyte	–	100 % eutro-mesophyte soils rich in calcium
Carici flavae—Cratoneuretum filicini Kovács et Felföldy 58	*Equisetum telmateia* *Pinguicula vulgaris* *Caltha palustris* ssp. *Laeta* *Cardamine amara* *Carex serotina* *Epilobium palustre* *Juncus effuses*	River mouth of the Suhard brook	Calcareous and permanently swampy substrate Under the form of biochores on the fossil trunks of the lake	43 % meso-hydrophyte 30 % hydrophyte 23 % mesophyte 2 % ultra-hydrophyte 2 % xerophyte	36 % circumpolar 32 % Euro-Asian 12 % European 8 % Central-European 12 % cosmopolite	46 % great variations 23 % weakly acido-neutrophile 19 % acido-neutrophile 10 % acidophile 2 % neutro-basiphile

(continued)

Table 10.3 (continued)

Association	Species	Place	Substratum	Hydrologic reaction	Phyto-geographic origin	Soil reaction
Angelico-Cirsietum oleracei Tx. 37	*Chaerophyllum hirsutum* *Galium palustre* *Mentha longifolia* *Equisetum palustre* *Lychnis flos cuculi* *Carex ovalis* *Cirsium palustre*	Wet or swampy meadows Banks of the Red Lake	Hydromorphic soils Gleyed soils	45 % meso-hydrophyte, 25 % mesophyte, 18 % hydrophyte, 5 % xero-mesophyte and 7 % amphy-tolerant	58 % Euro-Asian 25 % circumpolar 7 % Central-European 5 % European 5 % cosmopolite	55 % euriionic 23 % weakly acido-neutrophile 13 % acido-neutrophile 7 % acidophile 2 % neutro-basiphile
Potametum natantis Soó 27	*Potamogeton natans* *Lemna minor* *Callitriche palustris* *Myriophyllum spicatum* *Typha schuttleworthii* *Equisetum telmateja* *Glyceria plicata*	Shadowed bank	Alluvial soils	100 % ultra-hydrophyte	50 % circumpolar 50 % cosmopolite	–
Telekio-Petasitetum hybridi (Morariu 67) Resm. et Ratiu 74	*Chaerophyllum hirsutum* *Lamium maculatum* *Heracium sphondylium* *Galium aparine* *Agrostis capillaries* *Athyrium filix-femina*	Humid sector of the Licas valley Around Red Lake	Alluvial soils	49 % mesophile 40 % neutral 32 % meso-hydrophyte 13 % xero-mesophyte 4 % hydrophyte 2 % amphy-tolerant	49 % Euro-Asian 13 % Central European 11 % cosmopolite 9 % European 8 % circumpolar 4 % Carpathian-Balkan 2 % Mediterranean 2 % Arctic-Alpine 2 % endemic	5 % weakly acido-neutrophile 23 % acido-neutrophile 8 % acidophile 4 % neutro-basiphile

(continued)

Table 10.3 (continued)

Association	Species	Place	Substratum	Hydrologic reaction	Phyto-geographic origin	Soil reaction
Eleocharidetum palustris Sennikov 19	*Equisetum fluviatile* *Galium palustre* *Mentha aquatic* *Triglochin palustris* *Alisma plantago-aquatica* *Lycopus europaeus* *Stachys palustris* *Equisetum palustre*	Along the River Licas	Swampy soils	*Equisetum fluviatile* *Galium palustre* *Mentha aquatic* *Triglochin palustris* *Alisma plantago-aquatica* *Lycopus europaeus* *Stachys palustris* *Equisetum palustre*	48 % circumpolar 28 % Euro-Asian 16 % cosmopolite 8 % European	60 % large variations 32 % weakly acido-neutrophile 4 % acidophile 4 % acido-neutrophile

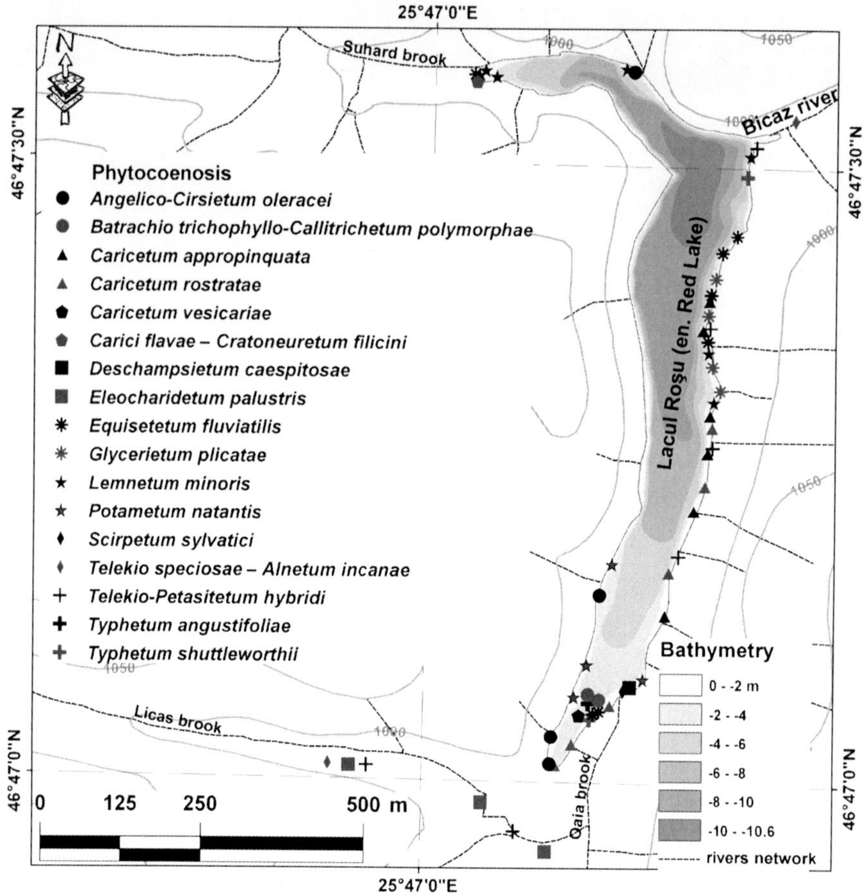

Fig. 10.10 Map of plants within the Red Lake area

occur in the tributaries of lake basins and conditions where wetlands appear: Suhard (Fig. 10.11), Licos, Vereschiu, Oaia (Oii), Calul, Sec, etc. Streams of an intermittent character show only small areas of wetlands and are not subject to this research.

The main stream, which feeds the lake in the southern sector, is the Oaia (Oii). In descending order, the Vereschiu and the Suhard rivers follow. These are streams that enter the lake basin directly. The other main river streams are only tributaries. Alluvia carried by these hydrographic arteries give rise to an alluvial cone (fan delta), which provides shelter, while the most extensive wetlands are associated with the lake.

The erosion of the surface in the hydrographic basin of the Red Lake is reduced. The compact forest contributes equally to the reduction. The surface of the catchment basin is 40.50 km^2, which is relatively small compared with the size of the lake. This has facilitated the low rate of silting of the lake and its durability.

Fig. 10.11 The wetland on the Suhard outlet

The large area contains solids. A slight increase in erosion was observed in the past 20 years in some forest areas. Privately owned, these were removed. However, the lake surface is being continuously reduced because the river arteries carry a large quantity of silt naturally. The transport of alluvia to the wetlands area increases and reduces the free water surface.

The Alpine climate specific to the zones located between 900 and 1,600 m means that precipitation is relatively high. The hydrologic regime of the Red Lake largely reflects the hydrologic regime of the river system tributaries. In this case, the lake water level has reduced oscillations. A slight increase is noted from melting snow and heavy rain over the long term. Slight decreases of level are registered in summer, due to a shortage of precipitation in winter, when rains are strong and water reserves of water remain entrenched in the snow. To maintain a balance between inputs and outputs, lake-basin water was banned in the entire basin. At the same time, the construction of other buildings was forbidden because they would increase water consumption and the balance installed may degrade.

Wetlands formed at the mouth of the main streams date from after the formation of the lake. Their existence is due to the outlet of alluvia in the lake water body. To reduce the rate of silting, dams were constructed on the Oaia and Vereschiu brooks and a trout farm on the Licas brook. This way, the rate of silting was significantly reduced in the Red Lake basin. However, the Administration of the Bicaz Gorges-Haghimas National Park is considering de-clogging the lake by removing the alluvial cone formed at the mouth of Oaia brook. This way, the lake surface will be increased and its life expectancy extended. The dredging will be in more turbulent waters but the community and mass of water bodies will suffer the consequences. The most important problem is the presence of wetlands, which have already been made official and form an integral part of the landscape. The dredging has been designed specifically to eliminate clogging them.

The Administration of the Bicaz Gorges-Haghimas National Park monitors tourism activity daily and ensures maintenance of ecologic behavior. In terms of good management of wetlands, the following measures are proposed: construction of another retention pond on the Oaia brook and a retention pond on the Suhard brook, halting deforestation and eliminating overgrazing in the catchment area; hydro-technical management of torrents, the cessation of business activities that are incompatible with National Park status, etc. The measures foreseen consider the preservation of wetlands and areas of biodiversity.

The Red Lake plays a very important role in the landscape and in tourism. With Lake Sfanta Ana (St Anne), it accounts for the largest annual number of tourists coming to visit an area of nature in the Eastern Carpathians. In terms of the transit of tourists it occupies, besides the Bicaz Gorges, it is the most popular place in Romania.

References

Barbu N, Lupaşcu G, Rusu C, Barbu A (1977) Solurile Munţilor Hăghimaş. Scientific Annals of the Alexandru Ioan Cuza University, Iaşi, sIIb, 23

Bendjoudi H, Fustec E (1996) Zones humides et régime des eaux. In: Fustec E, Frochot B (eds) Les fonctions et valeurs des zones humides. Université Paris VI et Dijon, Agence de l'Eau Seine-Normandie, France

Carter V (1996) Technical Aspects of wetlands. Wetland hydrology, water quality, and associated functions. In: National water summary on wetland resources, United States Geological survey, Water-Supply Paper 2425, Washington D.C

Devillers P, Devillers-Terschuren J, Ledant P (1991) CORINE-Biotopes manual. Habitats of the european community, Data specification Part 2. Commission of the european communities, Luxemburg

Dobrescu C, Ghenciu V (1970) Aspecte din vegetatia Lacului Rosu, Studii si comunicari. Mus Nat Sci, Bacau 3:129–136

Dobrescu C, Ghenciu V (1971a) Contribuţii la studiul biochoriilor din Lacul Roşu. Studies and Communications, Museum of Natural Sciences, Bacău

Dobrescu C, Ghenciu V (1971b) Contribuţii la cunoaşterea florei Lacului Roşu şi a împrejurimilor sale (Carpaţii Orientali). Annals of the Museum of Natural Sciences Piatra Neamţ, **2**

FICWD (1989) Federal manual for identifying and delineating jurisdictional wetlands. Federal Interagency Committee for Wetlands delineation, January 10, US Government Printing office, Washington DC

Goode DA, (1974) The significance of physical hydrology in the morphological classification of mires. Proc Int Peat Soc Symp, Glasgow (1973), I.P.S., Helsinki

Hurt GW, Carlisle VW (2001) Delineating hydric soils. In: Richardson JL, Vepraskas MJ (eds) Wetland Soils: genesis, hydrology, landscapes, and classification. Lewis Publishers, Boca Raton

Kentula ME (1996) Wetland restoration and creation. In: National water summary on wetland resources. United States Geological Survey, Water-Supply, 2425:87–92

Lefor MW, Kennard WC (1977) Inland wetland definition. Stors, Conn., University of Connecticut, Institute of Resources, Report 28

Lupaşcu A (1999) Studiu sinecologic comparativ in unele grupari vegetale higrofile din zona submontana a judetelor Suceava si Neamt. Corson Publishing House, Iasi

Nechita N (2003) Flora si vegetatia cormofitelor din masivul Hasmas, Cheile Bicazului si Lacului Rosu (The flora and vegetation of the cormophytes from Hasmas Massif, Bicaz Gorges, and the Red Lake). Bibliotheca Historiae Naturalis II, Natural Sciences Museum, Piatra Neamt, p 1–383

Report EPA (1983) Freshwater wetlands for wastewater management: environmental statement. Report EPA 904/9-83-107, Region IV USEPA, Atlanta

Report Y-87-1 (1987) Wetlands delineation manual. Environmental laboratory, US army engineer waterways experiment station, Vicksburg, MS, Rech. Report Y-87-1

Romanescu G, Gabriela Romanescu, Stoleriu C, Ursu A (2008) Inventarierea si tipologia zonelor umede si apelor adanci din Podisul Moldovei. Terra Nostra Publishing House, Iasi

Romanescu G (1996) Delta Dunării. Studiu morfohidrografic. Corson Publishing House, Iaşi

Romanescu G (2003) Hidrologie generală. Terra Nostra Publishing House, Iaşi, pp 1–574

Romanescu G, Cojocaru I (2010) Hydrogeological considerations on the western sector of the Danube Delta—a case study for the Caraorman and Saraturile fluvial-marine levees (with similarities for the Letea levee). Environ Eng Manage J 9(6):795–806

Romanescu G, Lupascu A, Stoleriu C, Raduianu D, Lesenciuc D, Vasiliniuc I, Romanescu G (2009) Inventarierea si tipologia zonelor umede si apelor adanci din grupa centrala a Carpatilor Orientali. "Al.I. Cuza" University Publishing House, Iasi

Romanescu G, Gabriela Romanescu, Minea I, Ursu A, Margarint MC, Stoleriu C (2005) Inventarierea si tipologia zonelor umede din Podisul Moldovei—Studiu de caz pentru judetele Iasi si Botosani. Didactical and Pedagogical Publishing House, Bucharest

Romanescu G, Gabriela Romanescu (2008) Inventarierea si tipologia zonelor umede si apelor adanci din Grupa Nordica a Carpatilor Orientali. Terra Nostra Publishing House, Iasi

Romanescu G, Lupaşcu A, Stoleriu C, Răduianu D, Lesenciuc D, Vasiliniuc I, Romanescu G (2010) Inventarierea şi tipologia zonelor umede şi apelor adânci din Carpaţii Orientali. Editura Terra Nostra, Iaşi

Smith RD (2001) Hydrogeomorphic approach to assessing wetland functions: Guidelines for developing Regional Guidebooks, ERDC/EL TR-01-29. US Army Engineer Research and Development Center, Vicksburg

Spencer WE (1994) Physiological response to flooding for wetland indicator plants. Technical Note VN-DL-1.1., US Army Engineer Waterways Experiment Station, Vicksburg, MS

Sprecher SW, Warne AG (2000) Accessing and using meteorological data to evaluate wetland hydrology. Technical Report ERDC/EL TR-WRAP-00-01, US Army Engineer Research and Development Center, Vicksburg, MS (http://www.wes.army.mil/el/wrap/pdf/wrap00-1/wrap00-1.pdf)

Tiner RW (1996) Technical aspects of wetlands. Wetland definitions and classifications in the United States, In: National water summary on wetland resources, United States Geological survey, Water-Supply Paper 2425, Washington D.C

Verger F (1968) Marais et wadden du littoral français. Etude de geomorphologie, Biscaye Frères, Bordeaux

Wentworth TR, Johnson GP, Kologiski RL (1988) Designation of wetlands by weighted averages of vegetation data: A preliminary evaluation. Water Resour Bull 24:389–396

Wilen BO, Carter V, Jones RJ (1996) Wetland management and research. Wetland mapping and inventory. In: National water summary on wetland resources, United States Geological survey, Water-Supply Paper 2425, Washington D.C

Chapter 11
Water Management

Abstract The Red Lake has started to be very influenced by human activity, due to the usage of water bodies for touristic purposes. From this point of view, there are several activities, such as deforestation, inappropriate tourist activities, pollution etc. that intervene in the natural erosion, transport and accumulation cycles, causing an increase of the silting rate, a phenomenon that will eventually lead to the disappearance of the Red Lake. For this reason, water management is a very important subject to bear in mind, when trying to adopt political or administrative decisions that involve landscape modifications in the drainage basin of the lake. The administration of the Red Lake—Bicaz Gorges Reservation monitors, on a daily basis, the tourist activity and ensures an ecological behavior. In order to conserve the hydrologic balance and to reduce clogging, the following actions are taken into account: catchment of deep underground waters and reducing the samplings directly from the tributary brooks of the Red Lake; stopping the deforestations of the hydrographic basin and eliminating over-pasturing. Other effective actions would be the hydro-technical management of the torrents and stopping the business activities incompatible with the National Park status; building a retention pond on the Suhard brook, etc. The measures foreseen consider the preservation of wetlands as areas of biodiversity and alluvium blockage.

Most of the lakes on Earth are strongly influenced by human activity. The biggest transformations occur around the water bodies used for water samplings and those with a high tourist potential (Battarbee et al. 2009; Bradbury et al. 2004; Goddard et al. 2008; Nolan and Brigham-Grettej 2007; Tarabay et al. 1991; Wetzel 2001). From this perspective, the lakes within the Eastern Carpathians are in an incipient use phase (Fig. 11.1). The conflicts between the landscape architects, the local organizations and the inhabitants will never be solved entirely (Christofides et al. 2005; Nkotagu 2008; Rahman and Wang 2010).

The Red Lake represents one of the most popular tourist locations within the Eastern Carpathians. This aquatic unit is very accessible, the reason for which it has a high tourist potential (Figs. 11.2, 11.3). The most important tributary streams

G. Romanescu et al., *Limnology of the Red Lake, Romania*,
DOI: 10.1007/978-94-007-6757-7_11,

Fig. 11.1 The two streams of the Red Lake: Oaia (*right*) and Suhard (*left*)

of the lacustrine basin are the following: Suhard, Licoş, Vereschiu, Oaia (Oii), Calul, Sec, Zgomotos, Hăghimaş and Hăghimaşul Negru. All these brooks are permanent. The other arteries have intermittent streams, most of them existing only during heavy rains.

The mountainous character of water streams often deteriorates the access roads. During the Communist Era, the Red Lake resort was really flourishing. After 1990, as consequence of the infrastructure lease, things grew worse. Only in the last few years, local tourism (but not international) seems to have revived. However, in reality, we are only talking about transit tourism.

The most important brook is Oaia (Oii), which supplies the lake in the southern sector, followed by the Vereschiu and Suhard brooks. They are also the brooks that open directly onto the lacustrine basin, the others being only tributaries of the main streams. According to the Horton-Strahler hierarchy, the tributaries of the Red

Fig. 11.2 Practicing leisure tourism in the area of the lacustrine basin

Fig. 11.3 Central setting of the Red Lake balneary and climatic resort

Lake are sixth-order. The surface of the catchment basin is 40.59 km^2, being relatively small compared to the dimensions of the lake. This very fact facilitates the low clogging rate and the lake endurance.

The maximum length, on the NW–SE direction, is 10.78 km and the average one is 3.76 km. The total length of the drainage network is 166 km with most arteries having an intermittent character. The watershed reaches altitudes exceeding 1,500–1,600 m. The maximum width of 5,900 m is found in the northern side of the basin. The elongated shape does not allow the accumulation of a large amount of water; thus, catastrophic floods are rare.

The climate of the Red Lake hydrographic basin is typically mountainous, with frequent thermal inversions, mostly in the winter. The icy bridge, up to 60 cm tick, is very long lasting. During the summer, the temperatures are relatively moderate, with low evapotranspiration. There has been a balance between the supply regime and that of water evacuation from the lacustrine basin, which also contributes to the conservation of the waters.

Mountain lakes do not have many fluctuations. Most level-related fluctuations occur in the case of plain lakes and those situated in areas with big rainfall fluctuations during a year (Altunkaynak et al. 2003). The hydrologic regime of the Red Lake reflects, largely, the hydrologic regime of the tributaries' hydrographic network. In this case, the water level within the lake does not oscillate much. Only slight increases are registered after the snow melts and during heavy, long-lasting rains. Also, there are slight decreases in the level during the summer and the winter (when the precipitations are solid and the water reserve remains cantoned in snow).

Fig. 11.4 Direct pollution sources around the Red Lake

The multi-annual average flow of the brooks opening onto the Red Lake basin is around 0.300–0.400 m^3/s. The multi-annual average flow going out of the lake is situated between 0.400 and 0.500 m^3/s. The 0.100 m^3/s difference in case of the lake mouth is given by the existence of karstic springs (emersed and submerse). In order to maintain a balance between the water inputs and outputs in the lacustrine basin, water catchment was prohibited for the entire hydrographic basin. At the same time, no other buildings operate, as they would increase the water consumption and the balance could be jeopardized.

Through the position that the Red Lake occupies within the hydrographic basin of Bicaz, the retention of alluvia within the lacustrine basin is facilitated. In this case, the waters situated upstream are fairer (Fig. 11.4). Because the lake is highly clogged, dams had to be built on the Oaia (Oii) brook and the Vereschiu brook (with big hydrographic basins) (Fig. 11.5), as well as a trout farm on the Licaş brook. This way, the clogging in the Red Lake basin was significantly reduced. Unfortunately, the two dams cannot deal with that much alluvium. The lake risks being even more clogged in the future because the aquatic vegetation of the two mouths keeps on developing (small depths favor its settlement) and the liquid or solid flow of the tributaries favors relatively high aggradation.

The lake and brook waters are not polluted because the few pensions and hotels built near the lake have septic tanks. At the same time, hunting and fishing are prohibited. Nonetheless, there are pensions and houses that spill uncontrollably and accidentally (on purpose) compost into the tributaries or even into the lake. This phenomenon can be perceived mainly during the night, because of the smell coming from the waters of the brooks (Figs. 11.6, 11.7).

The administration of the Red Lake—Bicaz Gorges Reservation monitors daily the tourist activity and ensures an ecological behavior. In order to conserve the

Fig. 11.5 The drainage lake dam on the Vereschiu brook (Piatra Roşie)

Fig. 11.6 Pollution sources after the passing of weekend tourists

hydrologic balance and to reduce clogging, the following actions are taken into account: catchment of deep underground waters and reducing the samplings directly from the tributary brooks of the Red Lake; stopping the deforestations of the hydrographic basin and eliminating over-pasturing. Other effective actions would be the hydro-technical management of the torrents and stopping the business activities incompatible with the National Park status; building a retention pond on the Suhard brook etc. The measures foreseen consider the preservation of wetlands as areas of biodiversity and alluvium blockage.

The unclogging of the lake, through measures to be applied in the future, in the two prograding sectors, is risky. The alluvia from the deposit will give a high degree of turbidity to waters. At the same time, the chemical products used in the basin area, deposited in the lacustrine basin, will be remobilized. The vegetation of the wetlands on the alluvial cones will be eliminated, the biodiversity being

Fig. 11.7 Direct pollution by spilling compost into the tributary brooks of the Red Lake

degraded. The sedimentary material eliminated must be deposited outside the hydrographic basin or it can be used as an amendment on the surrounding acid soils (high contents of calcareous-dolomitic material).

There has also been a proposition for a supra-elevation of the natural dam, with ripraps, in order to determine the elevation of the aquatic level (Grasu et al. 2010). This way, both the lacustrine surface and the depth can be increased. Unfortunately, the lake can no longer be considered natural but a weird hybrid. The elevation of the aquatic level will inevitably lead to the disappearance of fossil trees, the main attraction of the lake. Thus, the measure is utopian.

When talking about the vulnerability of the Red Lake, one should take into account that the lake water and its quality are not the only factors involved. Around the lake, a whole ecosystem was built, where numerous species of plants, amphibians, birds, fish and other types of inhabitants live. Thus, among the most important aspects with a disastrous effect upon the life of the lake should be mentioned: eutrophication, clogging, tourist activities (which also include pollution), deforestations, etc. (Fig. 11.8).

Eutrophication implies enriching the waters with nutritive substances, which leads to the development of vegetation and fauna, implicitly. This way, the vegetation has, first, a disastrous role by accelerating the organic clogging process, growing more and more in areas strongly affected by aggradation. This way, in the places where the vegetal layer creates adequate conditions, micro- and macro-faunas appear. During its lifespan, the Red Lake has passed through three successive phases: initially, in its first years, the regime was oligotrophic; the oligotrophic-mesotrophic phase, towards mesotrophic, when the macrophyte vegetation appeared; the current phase, characterized by a mesotrophic-eutrophic regime, as the lake goes through a continuous eutrophication process (Ghenciu 1968a, b).

Fig. 11.8 Road maintenance and water adduction works at the outlet of the Oaia (Oii) brook

This phenomenon is closely related to that of clogging, with the second eutrophic phase of the lake favoring hydrophilic vegetation. It accelerates the clogging process. Rivers represent the main clogging factor, as they carry an ever-growing amount of sediments. There are different types of alluvia, depending on their origin. There are numerous clayey alluvia, deposited in the central area of the lake but also crystalline sediments, brought from the south, after the desegregation of crystalline rocks. One can recognize them because of their high capacity of reflecting light and because of their high degree of roughness (Fig. 11.9).

Fig. 11.9 Crystalline sediments in the southern sector of the Red Lake

Besides natural clogging, an important amount of matter is also constituted by the compost thrown in by the local population or by tourists. There are two variants of pollution: the compost is thrown in from the banks (mostly from the pier area), or from the boats, during leisure walks on the lake.

Tourism may represent a disturbing factor for the natural state and the ecologic balance of the Red Lake (Bojoi and Paulencu 1975). Within the project elaborated by the City Hall of Gheorgheni buildings will be built with the roles of conserving the protected area of the lake and maintaining the ecological balance. Also within the project, the perturbing factors and the objectives affected by buildings or lake clogging were identified:

- destroying the underground shelters of the rodent mammals;
- destroying the holes, nests and underground galleries of the local fauna;
- altering the balance of the trophic chain, through stress caused by equipment;
- the noise created by equipment;
- destroying the local fauna and flora, because of equipment compacting the soil;
- the dust created by the building works affects the breathing system of the animals in the area;
- it is impossible to prevent certain accidents, etc.

Besides the various projects regarding the unclogging/clogging reduction/ landslide dam elevation, etc., the laws concerning the cleanness, fire ignition and fishing around the lake, etc., should also be enforced.

References

Altunkaynak A, Özger M, Sen Z (2003) Triple diagram model of level fluctuations in Lake Van. Turkey Hydrol Earth Syst Sci 7(2):235–244

Battarbee RW, Kernan M, Rose N (2009) Threatened and stressed mountain lakes of Europe: assessment and progress. Aquat Ecosyst Health Manage 12(2):118–128. doi:10.1080/14634980902905742

Bojoi I, Paulencu D (1975) Relieful carstic din regiunile muntoase Hăghimaş şi Rarău, element al potenţialului turistic. In: Proceedings of the national colloquium of applied geomorphology and geomorphologic mapping, Iaşi

Bradbury JP, Colman SM, Reynolds RL (2004) The history of recent limnological changes and human impact on Upper Klamath Lake, Oregon. J Paleolimnol 31:151–165

Christofides A, Efstratiadis A, Koutsoyiannis D, Sargentis GF, Hadjibiros K (2005) Resolving conflicting objectives in the management of the Plastiras Lake: can we quantify beauty? Hydrol Earth Syst Sci 9(5):507–515

Ghenciu IV (1968a) Regimul de oxigen al Lacului Rosu. Communications of the scientific session. Al.I.Cuza University, Iasi, pp 184–196

Ghenciu IV (1968b) Regimul termic al apei Lacului Rosu. Communications of the scientific session. Al.I.Cuza University, Iasi, pp 145–160

Goddard MA, Post CJ, English WR, Pike JW (2008) Examining the impacts of changing land use on biological integrity in streams using Geographical Information Systems and statistical modeling. Aquat Ecosyst Health Manage 11(2):230–242. doi:10.1080/14634980802111557

Grasu C, Miclăuş C, Brânzilă M, Baciu DS (2010) Munţii Hăşmaşului. Monografie geologic şi fizico-geografică. "Al.I.Cuza University" Publishing House, Iaşi

Nkotagu HH (2008) Lake Tanganyika ecosystem management strategies. Aquat Ecosyst Health Manage 11(1):36–41. doi:10.1080/14634980801891373

Nolan M, Brigham-Grettej J (2007) Basic hydrology, limnology, and meteorology of modern Lake El'gygytgyn Siberia. J Paleolimnol 37(1):17–35

Rahman HA, Wang C (2010) Preliminary approach to improve knowledge management in engineering management. Sci Res Essays 5(15):1950–1964

Tarabay AB, Villela RG, Espino GL (1991) Limnological aspects of a high-mountain lake in Mexico. Hydrobiologia, vol 224(1), Springer, pp 1–10

Wetzel R (2001) Limnology. Lake and River Ecosystems, (3rd edn). Academic Press, San Diego, CA

Chapter 12
Conclusions

Abstract There are relatively numerous standing waters in Romania but natural lakes, according to the universal definition, are rare. Most of the lakes in Romania are anthropic, being situated in all types of landforms, mainly in areas where there is a lack of water.

There are relatively numerous standing waters in Romania but natural lakes, according to the universal definition, are rare. Most of the lakes in Romania are anthropic, being situated in all types of landforms, mainly in areas where there is a lack of water. From this perspective, the northeastern part of Romania (the Moldavian Plain) is representative, where the first ponds were built, being mentioned in documents even since the reign of Stephen the Great (the 14th century). By the 18th century, there were over 3,500 ponds in the entire Moldavian Plateau. Nowadays, there are only 350 ponds, most of them situated in the relatively low landforms of the Moldavian Plateau, where there is a high scarcity of water (evapotranspiration—650–700 mm/year, precipitations—500–600 mm/year).

From a typological perspective, among the natural lakes can be: natural dam lakes (Red (Roşu), Cuejdu, Ştiol, etc.); volcanic (Saint Anne); fluviatile limans (Ostrov, Mârleanu, Vederoasa, etc.); lagoons (Razim, Sinoie, Siutghiol, etc.), of subsidence (Movila Miresii, Strachina, Amara, etc.); of deserted loops (Belciug, Erenciuc, etc.), etc.

The largest natural lake is Razim (415 km^2), part of the Razim-Sinoe lagoon complex (Razim, Sinoie, Goloviţa, and Zmeica). The deepest is the 27 m Zănoaga corrie lake. The best-known natural dam water body is the Red Lake (10.5 m deep), formed after a landslide that occurred in 1837. It is situated in the Hăghimaşul Mare Mountains, on the upper stream of the Bicaz River, upstream from the Bicaz Gorges. The deepest natural dam lake is Cuejdu, cu 16.20 m, formed in 1979, on the valley with the same way, in the Eastern Carpathians.

From a geologic perspective, the hydrographic basin of the Red Lake is situated in the Hăghimaşului (Hăşmaşului) syncline—Mesozoic. It is composed of limestones and Triassic dolomite, Lias-Dogger sandstones, Barremian-Albian conglomerates and Maastrichtian crystalline schists. As regards the soils, representative for this

G. Romanescu et al., *Limnology of the Red Lake, Romania*,
DOI: 10.1007/978-94-007-6757-7_12,

region are: chernozems, cambisols and mixed classes (rocks and rendzinas, prepodzols and podzols, dystric cambisols and eutric cambisols, rendzinas and subrendzina eutric cambisols). These soils are specific to pine and mixed woods.

The Hăghimaş Mountains landform has variations because of diversified geologic conditions and high tectonic fragmentation. From a genetic perspective, there are two hypotheses: it was formed after a landslide caused by heavy rains in 1837; or after a landslide generated by an earthquake of 1838. The first seems to be the most reliable.

The surface of the Red Lake catchment basin is 40.59 km^2. The maximum length of the basin is 10,783 m, with a 5,900 m length. The streams are short but with relatively big flows, typical to high mountain sectors with rich underground sources: Suhard, Licoş, Vereschiu, Oaia, Calul, Sec, Zgomotos, Hăghimaş and Hăghimaşul Negru. The other streams are intermittent, with relatively big slopes.

According to the Horton-Strahler hierarchy system, the tributaries of the Red Lake are sixth-order streams. As a percentage, the first-order streams are dominant. A detailed morphometric analysis of the hydrographic basin has been conducted, as it represents the whole, open system at a local level, one that receives and eliminates matter and energy. The physical-geographic parameters influence the ecologic elements.

Due to very modern equipment (ecodrill, topometric total station, high precision multiparameter, etc.), detailed measurements have been taken for the morphobathymetric elements and for the physical–chemical parameters of the water. All data were processed and then maps were generated in TNT Mips and ArcView.

The Red Lake basin is situated at an altitude between 966 m (upstream) and 965 m (downstream, at the outlet). It has a 12.01 ha surface, a 2,905.79 m perimeter, 1,361.72 m maximum length, 160.24 m maximum width and 10.5 m maximum depth. The biggest depths are registered at the submerged confluence of the two hydrographic arteries (Suhard and Oaia), behind the landslide dam. In the transverse profile, the left bank is more abrupt, having as a basis calcareous tough rocks, while the right bank is more even, corresponding to the landslide body.

A very interesting and important cartographic reproduction refers to the values of the slopes within the lacustrine basin. The living communities within the water body settle on favorable slopes, with low values and good light, while the high-value slopes are empty. It is an identical situation (at a reduced scale of the stream) with that of the emersed landform, with specific altitude and climatic layers.

The same category includes also the map concerning the slopes, with practical application in the distribution of vegetal and animal communities. From this perspective, the weakly illuminated slopes (northern, northeastern, northwestern) are few in number. In the longitudinal profile, certain thresholds were observed alongside the riverbed. The inside of the landslide dam is abrupt. The total water volume is 721,404.87 m^3, with different proportions on successive bathymetric layers, due to the conic shape of the lacustrine basin.

Pinewoods occupy most of the territory belonging to the hydrographic basin, which leads to a low erosion rate. The accelerated deforestations of the last period have determined the increase in the surface occupied by pastures. The largest deforested surfaces were registered between 1989 and 2001.

The distribution of organic carbon within the sediments and the quantity of humus within the limnic soils are high. Organic carbon reaches its highest value at 7.31 % and the humus—12.7 %. The highest values are found at the two extremities of the lacustrine basin (the mouths of the Oaia and Suhard brooks). The maximum clogging rate reached 1.5–1.6 m (2.8–3 cm/year) during 1957–2010. The climate of the area is specific to mountain units, with frequent thermal inversions, accompanied by the appearance of ice bridges during the winter (up to 60–70 cm thick).

The variations in the level of the lake have low values (60–70 cm), lower in the winter and higher during the spring, the rainy season and after the snow melts. The maximum water transparency value is 3.5 m, with means between 2 and 2.3 m. The transparency is extremely low during heavy rains, determining a solid transport, also through the streams.

The water reaches 18 °C in the summer and autumn and 4 °C in the spring. In the winter the ice bridge is formed. At over 10 m depth there is a constant temperature of 4–6 °C. At the surface, the pH is 7–8.2 and 7.6 at depths bigger than 10 m. The value of the dissolved oxygen depends upon temperature, dynamics and water quality. At the surface, the values are the highest—10.1 mg/L and at depths bigger than 8 m it goes down to near zero.

All physical–chemical parameters of the water have fluctuations depending on seasons and morpho-bathymetric layers. From a trophic point of view, the Red Lake is included in the meso-eutrophic (nutrients) and ultraoligotrophic (biologic) categories. Pines (spruce, silver fir, pin, larch, etc.) dominate the vegetation of the Red Lake hydrographic basin and mountain pastures tillaged with cereals (red fescue, matgrass, Lady's-slipper orchid, rosy vanilla orchid, etc.). The representative mammals are the following: brown bear, wild boar, red Carpathian deer, roe deer, etc. The blackbird, the sparrow hawk, the wood grouse, etc. represent the avifauna; it does not constitute an important site for birds of passages. Around and inside the lake, hydrophilic and hygrophilic vegetation is dominant. 59 taxons and 77 species of phytoplankton have been determined. The aquatic fauna comprises zooplankton, zoobenthos, fish and batrachians.

A special chapter was dedicated to the delimiting principles of the wetlands, depending on the interconnection of the ecologic parameters, among which slopes and their orientation are underlined. There are new principles to delimit wetlands within lacustrine basins. The repartition of the vegetal associations within the wetlands is highly related to the depth, the slopes and the orientation of the slopes.

In order to preserve the natural state of the lacustrine basin, a series of structural and non-structural measures have been taken. Some of them are viable but some others are, nonetheless, plain stupid. The unclogging of the lacustrine basin and the creation of alluvium traps on the main streams could solve the stability issue. Unfortunately, not all measures are welcome and applicable because there are not sufficient funds to start applying them.

The Red Lake has existed for a very, very long time. The current clogging rate is relatively low, hence the hope for a longer lifespan of the lake. The execution of works to reduce soil erosion is highly recommended but only inside the hydrographic basin, where the ecologic qualities of the lacustrine basin are not endangered. It should also be mentioned that any unclogging measure regarding the basin will inevitably lead to permanent alterations in the structure of the flora and fauna composition.

Index

G. Romanescu et al., *Limnology of the Red Lake, Romania*,
DOI: 10.1007/978-94-007-6757-7,

Printed by Publishers' Graphics LLC
LMO130731.15.16.297